Design of
ANALOG MULTIPLIERS
With Operational Amplifiers

Design of
ANALOG MULTIPLIERS
With Operational Amplifiers

K.C. Selvam
Department of Electrical Engineering
Indian Institute of Technology Madras
Chennai – 600036, India

CRC Press
Taylor & Francis Group
Boca Raton London New York

CRC Press is an imprint of the
Taylor & Francis Group, an **informa** business

CRC Press
Taylor & Francis Group
52 Vanderbilt Avenue
New York, NY 10017

First issued in paperback 2021

ISBN-13: 978-0-367-22894-1 (hbk)
ISBN-13: 978-1-03-209128-0 (pbk)

Library of Congress Cataloging-in-Publication Data

Names: Selvam, K. C., author.
Title: Design of analog multipliers with operational amplifiers / K.C. Selvam.
Description: Boca Raton : Taylor & Francis, a CRC title, part of the Taylor & Francis imprint, a member of the Taylor & Francis Group, the academic division of T&F Informa, plc, 2020. | Includes bibliographical references and index.
Identifiers: LCCN 2019015113| ISBN 9780367228941 (hardback : acid-free paper) | ISBN 9780429277450 (e-book)
Subjects: LCSH: Analog multipliers. | Operational amplifiers.
Classification: LCC TK7872.M8 S45 2020 | DDC 621.39/5--dc23
LC record available at https://lccn.loc.gov/2019015113

Visit the Taylor & Francis Web site at
http://www.taylorandfrancis.com

and the CRC Press Web site at
http://www.crcpress.com

Dedicated to my loving wife

S. Latha

Contents

Preface

When I was at Measurements and Instrumentation Laboratory, Electrical Engineering Department, IIT Madras, during the years 1988–1994, I worked on a project on wave energy. I needed a multiplier for my project. The multiplier circuit given by my project coordinator was very complex and expensive. I designed a simpler multiplier and showed it to Prof. Dr. V.G.K. Murti. He told me that the circuit was attractive and asked me to experiment with it. I experimented and showed him the test results and waveforms in a digital storage oscilloscope (DSO). He said, "Selvam excellent. Prepare a manuscript and publish it in IETE journals." In 1993, I published my first multiplier circuit in IETE student's journal. During the years 1993–1996, I published four analog multiplier circuits in Taylor & Francis Group journals. During the years 2011–2016, I published six analog multipliers in various national and international journals. During the years 2017–2018, I got concepts for 118 analog multiplier circuits, decided to publish a book and this is the result.

I am highly indebted to my

1. Mentor Prof. Dr. V.G.K. Murti who taught me how to get average value of a periodic waveform
2. Philosopher Prof. Dr. P. Sankaran who introduced me to IIT Madras
3. Teacher Prof. Dr. K. Radha Krishna Rao who taught me operational amplifiers
4. Director Prof. Dr. Bhaskar Ramamurti who motivated me to do this work
5. Gurunathar Prof. Dr. V. Jagadeesh Kumar who guided me in the proper ways of the scientific world
6. Trainer Dr. M. Kumaravel who trained me to do experiments with operational amplifiers
7. Encourager Prof. Dr. Enakshi Bhattacharya who encouraged me to get useful results
8. Leader Prof. Dr. Devendra Jalihal who kept me in a happy and peaceful official atmosphere
9. Supervisor Prof. Dr. K. Sridharan who supervised and monitored all my research work

I am indebted to Dr. S. Sathiyanathan, Dr. R. Rajkumar and Dr. V. Vasantha Jayaram who gave me medical treatment during the years 1996–2010. I am also indebted to Dr. Shiva Prakash, psychiatrist, and Dr. Saraswathi, general physician, who continue the medical treatment.

I am indebted to Prof. Dr. T.S. Rathore, former professor of IIT Bombay, who reviewed all my papers; and Prof. Dr. Raj Senani, former editor of *IETE Journal of Education*, who published most of my papers therein.

I thank my father Mr. Venkatappa Chinthambi Naidu, mother Mrs. C. Suseela, wife S. Latha, elder son S. Devakumar and younger son S. Jagadeesh Kumar for keeping me in a happy and peaceful residential atmosphere.

I thank my friend Dr. Bharath Bhikkaji who gave me a highly sophisticated DSO and asked me to take waveforms of my experimental setups in that DSO. This enabled me to easily verify my concepts experimentally. He also encouraged me throughout my research work.

I thank Dr. Gagandeep Singh, senior editor, Taylor & Francis/CRC Press, for showing keen interest in publishing all my concepts on function circuits. His hard work in producing this book is commendable.

I thank my assistants Mr. Nandakumar for all circuit and waveform drawings, Mrs. Padmapriya for word processing, Mr. B. Saravanan for circuit testing and Ms. Parvathy for proofreading.

Author

K.C. Selvam was born on 2 April 1968 in the Krishnagiri district of the state of Tamil Nadu, India. He obtained a diploma in electronics and communication engineering from Government Polytechnic College, Krishnagiri, Tamil Nadu, India in 1986. He graduated from the Institution of Electronics and Telecommunication Engineers, New Delhi, in 1994. He has done research and development work for the last 30 years and has published more than 34 research papers in various national and international journals. In 1996 he won the best paper award from IETE. He won the students' journal award of IETE in the year 2017. He developed India's first sleek and compact size transformerless multichannel autoscanning digital panel meters. At present he is working as a scientific staff member in the Department of Electrical Engineering, Indian Institute of Technology, Madras, India. His interests are in the design and development of function circuits to find their applications in modern measurements and instrumentation systems.

Useful Notations

V_1	First input voltage
V_2	Second input voltage
V_O	Output voltage
V_R	Reference voltage/peak value of first saw tooth waveform
V_T	Peak value of first triangular waveform
V_P	Peak value of second triangular wave/saw tooth wave
V_C	Comparator 1 output voltage in the first saw tooth/triangular wave generator
V_M	Comparator 2 output voltage by comparing saw tooth/triangular waves with one input voltage
V_N	Low pass filter input signal
V_{S1}	First generated saw tooth wave
V_{S2}	Second generated saw tooth wave
V_{T1}	First generated triangular wave
V_{T2}	Second generated triangular wave
V_S	Sampling pulse
V_1'	Slightly less than of V_1 voltage
$+V_{CC}$	Positive power supply
$-V_{CC}$	Negative power supply
$+V_{SAT}$	Positive op-amp saturation voltage
$-V_{SAT}$	Negative op-amp saturation voltage

Abbreviations

TDM	time division multiplier
MTDM	multiplexing time division multiplier
STDM	switching time division multiplier
PRM	peak responding multiplier
MPRM	multiplexing peak responding multiplier
SPRM	switching peak responding multiplier
PDM	peak detecting multiplier
MPDM	multiplexing peak detecting multiplier
SPDM	switching peak detecting multiplier
PSM	peak sampling multiplier
MPSM	multiplexing peak sampling multiplier
SPSM	switching peak sampling multiplier
PPRM	pulse position responding multiplier
PPDM	pulse position detecting multiplier
PPSM	pulse position sampling multiplier

Introduction

An analog multiplier is a function circuit that accepts two input voltages V_1 and V_2, and produces an output voltage $V_O = V_1 V_2 / K$, where K is the proportional constant and usually it will be a constant reference voltage V_R or peak value of reference clock V_T. The symbol of the multiplier is shown in Figure I.1. A multiplier that accepts inputs of either polarity is called a four-quadrant multiplier. A multiplier that accepts one bipolar input and other unipolar input is called a two-quadrant multiplier. A single-quadrant multiplier accepts both the inputs to be unipolar. Multiplier performance is specified in terms of accuracy and nonlinearity. Accuracy represents the maximum deviation of the actual output from the ideal value and this deviation is also referred to as the total error. Nonlinearity represents the maximum output deviation from the best-fit straight line for the case where one input is varied from the end while the other is kept fixed.

I.1 Characteristics

An ideal multiplier will have (1) infinite input impedance and (2) zero output impedance. But in practice no multiplier meets these characteristics 100% of the time. The input voltages have a finite differential and common mode voltage range with finite impedance. The output has a finite current output capacity and finite output impedance.

I.2 Specifications

The following are specifications of a multiplier:

1. Rated output (minimum voltage at rated current)
2. Output impedance
3. Maximum input voltage
    a. Maximum voltage for rated specification
    b. Maximum voltage for no damage to device
4. Input impedance
5. 3 dB bandwidth
6. Output slew rate
7. Output settling time

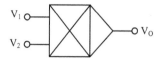

FIGURE I.1
Multiplier symbol.

I.3 Errors

The actual output of a multiplier can be expressed as

$$V_o = (k_m + \Delta k_m)(V_1 + V_{1IO})(V_2 + V_{2IO}) + V_{oo} + V_1 f + V_2 f + V_{n(1,2)} \qquad \text{(I.1)}$$

where

Δk_m = Scale factor error
V_{1IO} = Input offset voltage of V_1 input
V_{2IO} = Input offset voltage of V_2 input
V_{oo} = Output offset voltage
$V_1 f$ = Nonlinear feedthrough from V_1 input to V_o output
$V_2 f$ = Nonlinear feedthrough from V_2 input to V_o output
$V_n(1,2)$ = non linearity in gain response

Rearrange Equation I.1 given

$$V_o = K_m v_1 + v_2 + \Delta_m(v_1 + v_2) + \left[(K_m + \Delta k_m)_{V1V2I0+V1f}\right]$$

$$+ \left[(k_m + \Delta k_m)(V_2 V_{1I0}) + (V_{00} + V_n)_{(1,2)}\right]$$

$K_m v_{1+} v_2 \rightarrow$ ideal output
$\Delta_m(v_{1+} v_2) \rightarrow$ gain error
$(K_m + \Delta k_m)_{V1V2I0+V1f} \rightarrow$ total V_1 input feedthrough
$(k_m + \Delta k_m)(V_2 V_{1I0}) \rightarrow$ total V_2 input feedthrough
$V_{oo} \rightarrow$ output offset
$(V_{00} + V_n)_{(1,2)} \rightarrow$ nonlinear output

The multiplier has four external adjustments:

1. The V_1 input offset null (for V_{1Io})
2. The V_2 input offset null (for V_{2Io})
3. The output offset null (for V_{oo})
4. A gain adjustment (for V_n (1,2))

I.4 Individual Error

When any one of the inputs of a multiplier is zero, the output is supposed to be zero. But in practical multipliers, the following errors will occur: (1) output offset voltage; (2) V_1 input feedthrough, which is a small error feeding to the output from V_1 input when the V_2 input is zero; and (3) V_2 input feedthrough, which is a small error feeding to the output from V_2 input when V_1 input is zero. Output offset voltage is the output voltage when both input voltages V_1 and V_2 are zero. This can be nulled with suitable offset adjustment presets in the op-amps used for multiplication. The V_1 input feedthrough has a V_1 input signal being multiplied by finite V_2 input. When V_2 input is at zero, output should be at zero volts. However, there will be some offset voltage at the output. This offset can be adjusted with a preset to zero voltage. This is called "nulled." The feedthrough error can be minimized by nulling the input offset voltages of both input voltages. The feedthrough error will increase due to input offset temperature drifts and hence the multiplier should have self-nulling facilities. The feedthrough error also depends on frequency, and the error will increase if frequency is increased.

There may be a difference between V_1 input through and V_2 input through errors. The feedthrough can be expressed as

$$\text{Feedthrough} = \frac{\text{Peak to peak output voltage}}{\text{Either } V_1 \text{ or } V_2 \text{ input at zero}}$$

The typical value of feedthrough is 50 mV (0.5%) or less.

Gain error is the output error when all input voltages V_1 and V_2 are at maximum values (let us say $V_1 = V_2 = 10\ V_{max}$) for both the output offset and input offset and nulled.

Nonlinearity error is the maximum deviation of the input/output responses to the ideal characteristic. Typical commercial multipliers have a nonlinearity error between 0.01% and 0.5%.

There are two dynamic errors: (1) distortion, which increases at higher frequencies; and (2) bandwidth, which depend on the dc signal levels.

I.5 Offset Nulling

The following steps are to be followed for offset nulling in a multiplier:

1. Connect the V_1 input to the ground potential (GND). Apply low frequency sine wave of maximum amplitude to V_2 input (e.g., 10 V max at 50 Hz). Null the V_1 offset until the output becomes zero.

2. Connect the V_2 input to the GND. Apply a low frequency sine wave of maximum amplitude to the V_1 input. Null the V_2 offset until the output becomes zero.

3. Connect both the V_1 input and V_2 input to the GND. Null the output offset for zero output voltage.

I.6 Multiplier Types

Log–antilog multipliers, field-effect transistor (FET) multipliers, transconductance multipliers, Gilbert multiplier cells, and triangle averaging multipliers are few examples of earlier multipliers. The author has proposed many analog multiplier circuits in various national and international journals [3–10]. They are commonly classified as time division multipliers, peak responding multipliers and pulse position responding multipliers. Peak responding multipliers are further classified as peak detecting multipliers and peak sampling multipliers. Pulse position responding multipliers are classified as pulse position detecting multipliers and pulse position sampled multipliers. Saw tooth wave–based multipliers and triangular wave–based multipliers are a few examples of time division multipliers. Double single-slope peak detecting, double dual-slope peak detecting and pulse width integrated peak detecting multipliers are a few examples for peak detecting multipliers. Double single-slope sampling, double dual-slope sampling, pulse width integrated sampling and pulse position sampling multipliers are a few examples for peak sampling multipliers. All these types of multipliers are described in this book with detailed explanations.

1

Basic Concepts and Components

The inverting amplifier, non-inverting amplifier, integrator, comparator, low pass filter, analog switch, analog multiplexers, astable multivibrator, peak detector, and sample and hold circuits are the basic components of a multiplier. All multipliers are developed using the aforementioned components. These components are briefly discussed in this chapter.

1.1 Inverting Amplifier

Figure 1.1 shows an inverting amplifier using an op-amp. Since the non-inverting terminal (+) of the op-amp is grounded through the resistor R_P, the voltage V_A at the non-inverting terminal (+) of the op-amp is zero volts. The op-amp is in negative closed loop feedback and hence its non-inverting terminal voltage will be equal to its inverting terminal voltage, i.e.,

$$V_A = V_B = 0$$

The current through resistor R_1 will be

$$I = \frac{V_I - V_B}{R_1} = \frac{V_I}{R_1} \tag{1.1}$$

Since the op-amp has infinite input impedance, the current (I) does not enter into the op-amp and flows through resistor R_2. The voltage across resistor R_2 will be

$$V_{R2} = IR_2 = \frac{V_I}{R_1} R_2 \tag{1.2}$$

The negative feedback forces the op-amp to produce an output voltage that maintains a virtual ground at the op-amp inverting input. The output voltage is given as

$$V_O = V_B - V_{R2} = -V_{R2}$$

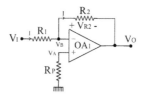

FIGURE 1.1
Inverting amplifier.

$$V_O = -\left(\frac{R_2}{R_1}\right)V_I \tag{1.3}$$

If $R_1 = R_2$, then

$$V_O = -V_I \tag{1.4}$$

1.2 Non-Inverting Amplifiers

Figure 1.2 shows a non-inverting amplifier, type I. The op-amp is at negative closed loop feedback and its inverting terminal (−) voltage will equal its non-inverting terminal (+) voltage, i.e.,

$$V_A = V_B = V_I$$

The current through resistor R_1 will be

$$I = \frac{V_B}{R_1} = \frac{V_I}{R_1} \tag{1.5}$$

The current (I) comes from resistor R_2 and it does not enter op-amp, as the op-amp has infinite input impedance. The voltage across feedback resistor R_2 will be

$$V_{R2} = IR_2 = \frac{V_I}{R_1}R_2 \tag{1.6}$$

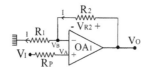

FIGURE 1.2
Non-inverting amplifier.

The output voltage is given as

$$V_O = V_B + V_{R2}$$

$$V_O = V_I + \left(\frac{R_2}{R_1} V_I \right)$$

$$V_O = V_I \left(1 + \frac{R_2}{R_1} \right) \tag{1.7}$$

If $R_1 = R_2$, then

$$V_O = 2V_I \tag{1.8}$$

Figure 1.3 shows a unity gain non-inverting amplifier, type II. Let us analyze the circuit using the superposition principle.

First the non-inverting terminal (+) is grounded through resistor R_P and the input voltage V_I is given to the inverting terminal through resistor R_1. As discussed in Section 1.1, the circuit will work as an inverting amplifier and if $R_1 = R_2$, the output voltage will be

$$V_{O1} = -V_I \tag{1.9}$$

Next, the inverting terminal (−) is grounded through resistor R_1 and the input voltage is given to the non-inverting terminal (+) through resistor R_P. As discussed earlier, the circuit will work as a non-inverting amplifier and if $R_1 = R_2$, the output voltage will be

$$V_{O2} = 2V_I \tag{1.10}$$

By the superposition principle the actual output voltage will be the addition of Equations 1.9 and 1.10:

$$V_O = V_{O1} + V_{O2} \tag{1.11}$$

$$V_O = V_I \tag{1.12}$$

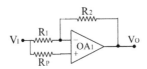

FIGURE 1.3
Unity gain non-inverting amplifier.

1.3 Integrator

Figure 1.4a shows an integrator using an op-amp. Since the non-inverting terminal (+) is at ground potential, the inverting terminal (−) will also be at ground potential by a virtual ground. The current through the resistor R will be

$$I = \frac{V_I - 0}{R} = \frac{V_I}{R} \tag{1.13}$$

Due to the op-amp's high input impedance, the current I will not enter into the op-amp and flows through the capacitor C. The voltage across capacitor C will be

$$V_C = \frac{q}{C} \tag{1.14}$$

where, q is the charge that exists on plates of the capacitor. The relation between the current (i) and charge (q) is given as

$$i = \frac{dq}{dt}$$

$$q = \int I dt \tag{1.15}$$

Equation 1.15 in Equation 1.14 gives

$$V_C = \frac{\int I dt}{C} \tag{1.16}$$

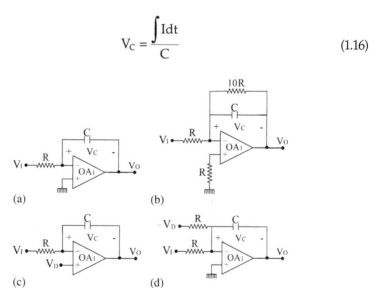

FIGURE 1.4
(a) Integrator. (b) Practical integrator. (c) Differential integrator. (d) Equivalent circuit of part (c).

The negative feedback forces the op-amp to produce an output voltage that maintains a virtual ground at the op-amp inverting input. The output voltage is given as

$$V_O = 0 - V_C = -V_C \qquad (1.17)$$

Equation 1.16 in Equation 1.17 gives

$$V_O = -\frac{1}{C}\int I dt \qquad (1.18)$$

Equation 1.13 in Equation 1.18 gives

$$V_O = -\frac{1}{RC}\int V_I dt = -\frac{V_I}{RC}t \qquad (1.19)$$

Figure 1.4b shows the circuit of a practical integrator. If a square wave is given as the input to the integrator, then a triangular wave will be the output of the integrator.

Figure 1.4c shows a differential integrator. Its output is given as

$$V_O = \frac{1}{RC}\int (V_D - V_I)dt = \frac{(V_D - V_I)}{RC}t \qquad (1.20)$$

1.4 Comparator

A comparator has two input terminals and one output terminal as shown in Figure 1.5.

The non-inverting terminal (+) and inverting terminal (−) are the two input terminals. If the non-inverting terminal voltage is greater than the inverting terminal voltage, then the output goes to HIGH (positive saturation +V_{SAT}). If the non-inverting terminal voltage is lower than the inverting terminal voltage, then the output goes to LOW (negative saturation −V_{SAT}).

Comparison of saw tooth waveform. A saw tooth wave V_{S1} of peak value V_R and time period T is compared with a voltage V_1 by a comparator as shown in Figure 1.6.

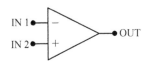

FIGURE 1.5
Comparator.

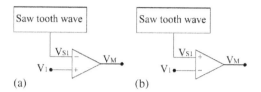

(a)   (b)

FIGURE 1.6
(a) Saw tooth comparison (case I). (b) Saw tooth comparison (case II).

In case I, a saw tooth wave is given to the inverting terminal of the comparator (–) and the input voltage is given to the non-inverting terminal (+) of the comparator.

In case II, a saw tooth wave is given to the non-inverting terminal (+) of the comparator and the input voltage is given to the inverting terminal (–) of the comparator.

In both cases a rectangular wave V_M is generated at the output of the comparator as shown in Figure 1.7. The ON time (case I) or OFF time (case II) of the generated rectangular waveform V_M will be

$$\delta_T = \frac{\text{Input voltage}}{\text{Peak value of saw tooth wave}} \text{Time period} = \frac{V_1}{V_R} T \quad\quad (1.21)$$

Comparison of triangular waveform. A triangular wave of peak value V_T and time period T is compared with a voltage V_1 by a comparator as shown in Figure 1.8. A rectangular waveform V_M is generated at the output of the comparator as shown in Figure 1.9.

In case I, the triangular wave is given to the inverting terminal (–) of the comparator and the input voltage is given to the non-inverting terminal (+) of the comparator.

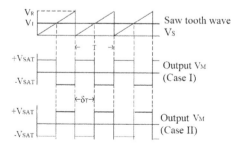

FIGURE 1.7
Associated waveforms of Figure 1.6.

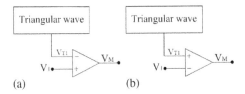

FIGURE 1.8
(a) Comparison of triangular wave (case I). (b) Comparison of triangular wave (case II).

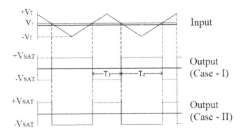

FIGURE 1.9
Associated waveforms of Figure 1.8.

In case II, the triangular wave is given to the non-inverting terminal of the comparator and input voltage is given to the inverting terminal of the comparator.

The OFF time (case I) or ON time (case II) T_1 of this rectangular wave V_M is given as

$$T_1 = \frac{\text{Peak value of triangular wave} - \text{Input voltage}}{\text{Twice the peak value of triangular wave}} \text{Time period}$$

(1.22)

$$= \frac{V_T - V_1}{2V_T} T$$

The ON time (case I) or OFF time (case II) T_2 of this rectangular waveform V_M is given as

$$T_2 = \frac{\text{Peak value of triangular wave} + \text{Input voltage}}{\text{Twice the peak value of triangular wave}} \text{Time period}$$

(1.23)

$$= \frac{V_2 + V_1}{2V_T} T$$

Time period $T = T_1 + T_2$

(1.24)

1.5 Low Pass Filter

Figure 1.10a shows a simple RC low pass filter and Figure 1.10b shows a first-order op-amp low pass filter.

In case I, a periodic rectangular pulse waveform V_N of peak value V_2, ON time δ_T, and time period T as shown in Figure 1.11a is given to a low pass filter. The output of the low pass filter is the average value of the waveform V_N and is given as

$$V_{avg} = \frac{1}{T}\int_0^{\delta_T} V_2 dt \qquad (1.25)$$

$$V_{avg} = \frac{V_2}{T}\delta_T \qquad (1.26)$$

In case II, a rectangular wave V_N with $\pm V_2$ as the peak-to-peak values and time period T as shown in Figure 1.11b is given to a low pass filter. The low pass filter output is the average value of V_N and is given as

$$V_{avg} = \frac{1}{T}\left[\int_O^{T_2}(+V_2)dt + \int_{T_2}^{T_1+T_2}(-V_2)\,dt\right] \qquad (1.27)$$

$$V_{avg} = \frac{1}{T}\left[V_2T_2 - V_2T_1 - V_2T_2 + V_2T_2\right]$$

$$V_{avg} = \frac{V_2}{T}(T_2 - T_1) \qquad (1.28)$$

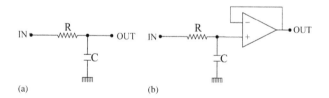

(a)   (b)

FIGURE 1.10
(a) Simple low pass filter. (b) Op-amp low pass filter (first order).

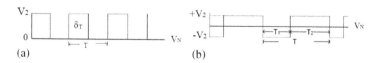

(a)   (b)

FIGURE 1.11
(a) Rectangular pulse waveform. (b) Rectangular wave.

1.6 Analog Switches

The symbol for an analog switch is shown in Figure 1.12. It has three terminals: CON, IN/OUT and OUT/IN. If the control (CON) pin is LOW, the switch is opened so that the IN/OUT and OUT/IN terminals are disconnected. If the control (CON) pin is HIGH, the switch is closed so that the IN/OUT and OUT/IN terminals are connected.

Analog switches are available in an integrated circuit (IC) package of CMOS CD4066 IC. The pin details of this CD4066 IC are given in the appendix.

Figure 1.13a shows a transistor series switch. If the control input V_M is LOW and the transistor Q is OFF, then the collector voltage will not exist on the emitter voltage:

$$V_N \sim 0$$

If the control input V_M is HIGH and the transistor Q is ON, then the collector voltage will exist on the emitter terminal:

$$V_N \sim V_2$$

Figure 1.13b shows a transistor shunt switch. If the control input V_M is LOW and transistor Q is OFF, then $V_N \sim V_2$ (R_C value is very low). If the control input V_M is HIGH and transistor Q is ON, then $V_N \sim 0$.

Figure 1.14a shows a field-effect transistor (FET) as a series switch. If the control input is HIGH (+Vcc), zero volts will exist on the gate terminal; the FET is ON and acts as a closed switch: OUT ~ IN. If the control input CON is

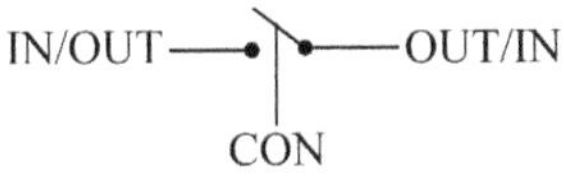

FIGURE 1.12
Switch symbol.

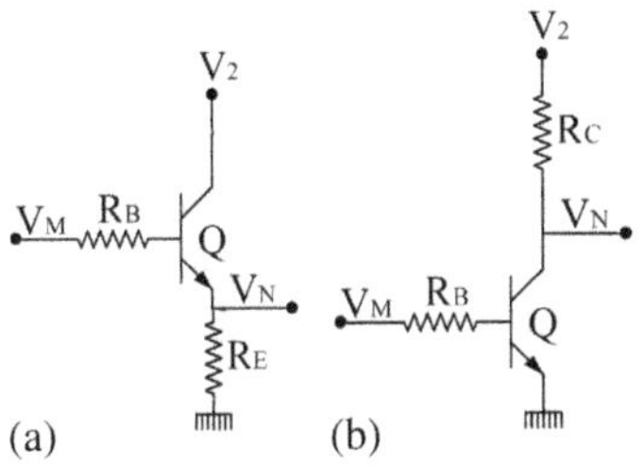

FIGURE 1.13
(a) Transistor series switch. (b) Transistor shunt switch.

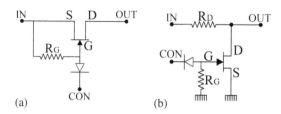

FIGURE 1.14
(a) FET series switch. (b) FET shunt switch.

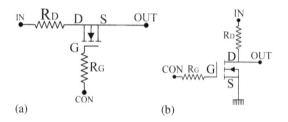

FIGURE 1.15
(a) MOSFET series switch. (b) MOSFET shunt switch.

LOW(–Vcc), then negative voltage will exist on gate terminal, the FET is OFF and acts as an open switch: OUT ~ 0.

Figure 1.14b shows a FET shunt switching circuit. If the control input CON is HIGH (+V_{CC}), then zero volts will exist on gate; FET is ON and zero volts will be output: OUT ~ 0. If the control input CON is LOW (–V_{CC}), then negative voltage will exist on the gate terminal; the FET operated on the cut-off region and acts as an open circuit. The output will be OUT ~ IN.

Figure 1.15a shows a metal-oxide-semiconductor field-effect transistor (MOSFET) series switch. If the control input is HIGH (+V_{DD}), the channel resistance becomes so small and allows maximum drain current to flow. This is the saturation mode and the MOSFET is completely ON and acts as a closed circuit: OUT ~ IN. If the control input is LOW (V_{SS}), the channel resistance becomes HIGH and no current flows from the drain. This is the cut-off region and MOSFET is completely OFF and acts as an open switch: OUT ~ 0.

Figure 1.15b shows a MOSFET shunt switch. If the control input is HIGH (+V_{DD}), the channel resistance becomes small and allows maximum drain current to flow. This is the saturation mode and the MOSFET is completely ON and acts as a closed circuit: OUT ~ 0. If the control input is LOW (V_{SS}), the channel resistance becomes HIGH and no current flows from the drain. This is the cut-off region and MOSFET is completely OFF and acts as an open switch: OUT ~ IN.

1.7 Analog Multiplexers

Figure 1.16 shows the symbol for an analog triple 2-to-1 multiplexer. Each multiplexer has four terminals. In case of multiplexer M_1, it has 'ay', 'ax', 'a' and 'A' terminals. In case of multiplexer M_2, it has 'by', 'bx', 'b' and 'B' terminals. In case of multiplexer M_3, it has 'cy', 'cx', 'c' and 'C' terminals.

In multiplexer M_1, if pin 'A' is HIGH, then 'ay' is connected to 'a' and if pin 'A' is LOW, then 'ax' is connected to 'a.' In multiplexer M_2, if the pin 'B' is HIGH, then 'by' is connected to 'b' and if the pin 'B' is LOW, then 'bx' is connected to 'b'. In multiplexer M_3, if the pin 'C' is HIGH, then 'cy' is connected to 'c' and if the pin 'C' is LOW, then 'cx' is connected to 'c'.

All the three multiplexers M_1, M_2 and M_3 are available in one IC package of CMOS CD4053 IC. The pin details of this CD4053 IC are given in the appendix.

Figure 1.17 shows an analog multiplexer using transistors. If the control input V_M is HIGH ($+V_{CC}$), transistor Q_1 is ON, Q_2 is OFF and $+V$ will appear at V_N. If control input CON is LOW ($-V_{CC}$), transistor Q_1 is OFF, Q_2 is ON and $-V$ will appear at V_N.

Figure 1.18 shows an analog multiplexer using FETs. If the control input CON is HIGH ($+V_{CC}$), FET Q_1 is ON, FET Q_2 is OFF and $+V$ will appear at V_N. If control input CON is LOW ($-V_{CC}$), FET Q_1 is OFF, FET Q_2 is ON and $-V$ will appear at V_N.

Figure 1.19 shows an analog multiplexer using MOSFETs. If the control input CON is HIGH ($+V_{CC}$), MOSFET Q_1 is ON, MOSFET Q_2 is OFF and $+V$ will appear at V_N. If control input CON is LOW ($-V_{CC}$), MOSFET Q_1 is OFF, MOSFET Q_2 is ON and ($-V$) will appear at V_N.

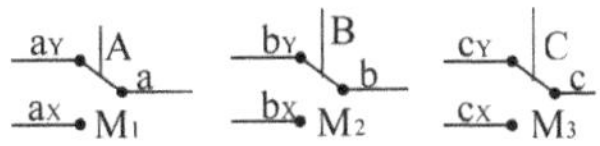

FIGURE 1.16
Triple 2-to-1 multiplexers.

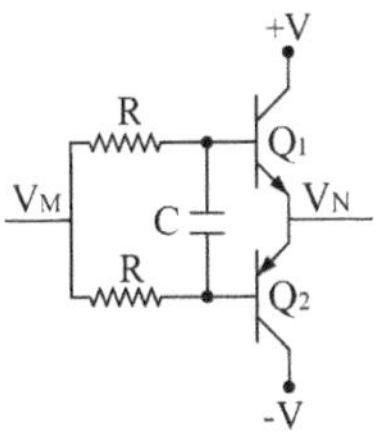

FIGURE 1.17
Transistor multiplexer.

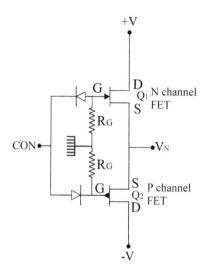

FIGURE 1.18
FET multiplexer.

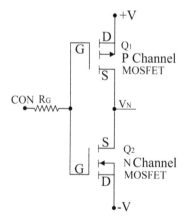

FIGURE 1.19
MOSFET multiplexer.

1.8 Peak Detector

The peak detector detects the maximum value of a signal over a period of time. Figure 1.20a shows a simple diode–capacitor peak detector.

The capacitor C is charged by the input signal through the diode. When the input signal falls, the diode is reverse biased and the capacitor voltage retains the peak value of the input signal. This simple circuit has errors

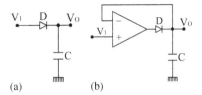

FIGURE 1.20
(a) Simple diode–capacitor peak detector. (b) Op-amp peak detector.

because of the diode forward voltage drop. These forward voltage drop errors can be removed by replacing the diode with a precise diode as shown in Figure 1.20b. It operates in either a peak tracking mode or a peak storage mode. During the peak tracking mode, the peak detector tracks the input toward a peak value. During the peak storage mode, the peak detector is held constant with the peak value. Figure 1.20 shows positive peak detectors and if we interchange the polarity of diode D, then they will become negative peak detectors.

If a saw tooth wave of peak value V_P is given to a peak detector, the peak detector output is a direct (dc) voltage of V_P. Similarly if a triangular wave of $\pm V_P$ value is given to a peak detector and if the peak detector is positive, then its output will be a dc voltage of $+V_P$. If the peak detector is negative, then its output will be a dc voltage of $-V_P$.

1.9 Sample and Hold Circuit

The sample and hold circuit samples and holds the input signal at a particular instant to determine the sampling pulse. Figure 1.21a shows a simple sample and hold circuit.

Let a saw tooth wave of peak value V_P and time period T be given as input to the sample and hold circuit. As shown in Figure 1.21a, a sampling pulse V_S is given to the sample and hold circuit. The switch S_1 is closed during the HIGH time of sampling pulse V_S and at that particular time of the input signal is given to the capacitor C and the capacitor C holds this signal even

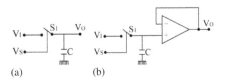

FIGURE 1.21
(a) Simple sample and hold circuit. (b) Op-amp sample and hold circuit.

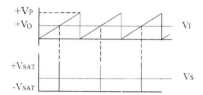

FIGURE 1.22
Waveforms of Figure 1.21.

if the switch S_1 opens during LOW time of the sampling pulse V_S. Hence the instant value of the input signal is sampled and held with the sampling pulse V_S. This is illustrated graphically in Figure 1.22. An op-amp buffer can be added at the output in order to avoid the loading effect, as shown in Figure 1.21b.

1.10 Astable Multivibrator

Figure 1.23 shows an astable multivibrator using an op-amp. Let us assume initially the op-amp output is LOW (i.e., negative saturation). The voltage at the non-inverting terminal will be

$$V_A = \beta(-V_{SAT})$$

$$\beta = \frac{R_1}{R_1 + R_2} \qquad (1.29)$$

The voltage at the inverting terminal V_B will be positive with regard to V_A and its potential is decreasing, i.e., C_1 charges down through R_3. When the potential difference between the two input terminals approaches zero, the op-amp comes out of saturation. The positive feedback from the output to

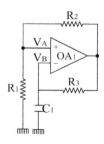

FIGURE 1.23
Astable multivibrator.

terminal V_A causes regenerative switching, which drives the op-amp to positive saturation. Capacitor C_1 charges up through R_3 and V_B potential rises exponentially; when it reaches $V_B = \beta(+Vcc)$ the circuit switches back to the state in which the op-amp is in negative saturation. The sequence therefore repeats to produce the square waveform of time period T at its output. The time period T is given as

$$T = 2R_3C_1 \ln\left(1 + 2\frac{R_1}{R_2}\right) \tag{1.30}$$

Voltage-to-period converter. If in the astable multivibrator shown in Figure 1.23 the R_2 terminal is removed from the output terminal and a controller is added between R_2 and output as shown in Figure 1.24, then the circuit will work as a voltage-to-time period converter. The time period T is given as

$$T = V_C K_1 \tag{1.31}$$

where K_1 is constant depends on Equation 1.30 and the op-amp saturation voltage or supply voltage V_{CC}.

Voltage-to-frequency converter. If in the astable multivibrator shown in Figure 1.23 the R_3 terminal is removed from the output terminal and a controller is added between R_3 and output as shown in Figure 1.25, then the circuit will work as a voltage-to-frequency converter. The frequency f is given as

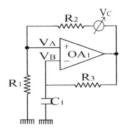

FIGURE 1.24
Voltage-to-period converter.

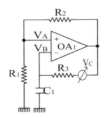

FIGURE 1.25
Voltage-to-frequency converter.

$$f = V_C K_1 \qquad (1.32)$$

where K_1 is constant depends on Equation 1.30 and the op-amp saturation voltage or supply voltage V_{CC}.

Tutorial Exercises

1.1 −5 V is required to be obtained from a 1 V source. Design a suitable op-amp circuit.

1.2 Find the output voltage in the circuit shown in Figure 1.26.

1.3 Find the output voltage in the circuit shown in Figure 1.27.

1.4 For an integrator, −1 V is applied at t=0. Determine the resistor–capacitor (RC) time constant required such that the output reaches +10 V at t=1 mS.

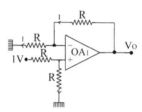

FIGURE 1.26
Circuit for Tutorial Exercise 1.2.

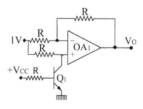

FIGURE 1.27
Circuit for Tutorial Exercise 1.3.

2

Conventional Multipliers

Log–antilog multipliers, resistor multipliers using field-effect transistors (FETs) as the voltage variable, variable transconductance multipliers, Gilbert multiplier cells, triangle wave averaging multipliers and quarter–square multipliers are conventional multipliers and are described in this chapter.

2.1 Log–Antilog Multiplier

Figure 2.1 shows the block diagram of a multiplier using log–antilog amplifiers. From Figure 2.1, the outputs of log amplifiers OA_1 and OA_2 are given as

$$V_X = V_T \ln \frac{V_1}{V_R} \tag{2.1}$$

$$V_Y = V_T \ln \frac{V_2}{V_R} \tag{2.2}$$

In Equations 2.1 and 2.2 V_T and V_R are constants of log amplifiers. The output of the unity gain summer OA_3 is

$$V_Z = V_X + V_Y$$

$$V_Z = V_T \ln \left(\frac{V_1 V_2}{V_R^2} \right) \tag{2.3}$$

The output of the unity gain summer V_Z is given to the antilog amplifier. The output voltage V_O is given as

$$V_O = V_R \ln^{-1} \left[\frac{V_Z}{V_T} \right]$$

$$V_O = V_R \ln^{-1 \ln \left(\frac{V_1 V_2}{V_R^2} \right)}$$

$$V_O = \frac{V_1 V_2}{V_R} \tag{2.4}$$

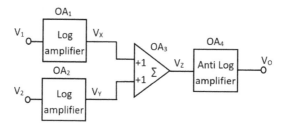

FIGURE 2.1
Block diagram of a log–antilog multiplier.

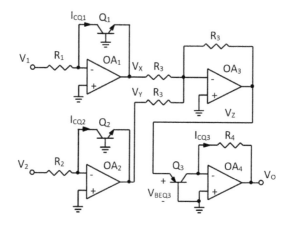

FIGURE 2.2
Circuit diagram of a log–antilog multiplier.

Figure 2.2 shows the complete circuit diagram of a log–antilog multiplier using op-amps and transistors. The logarithmic operation of the bipolar transistor is

$$V_{BE} = V_T \ln \frac{I_C}{I_S}$$

where V_{BE} is the base emitter voltage, $V_T = KT/Q$, I_C is the collector current and I_S is the emitter saturation current. In Figure 2.2 the collector current of transistor Q_1 will be

$$I_{CQ1} = \frac{V_1}{R_1} \qquad (2.5)$$

The output of log amplifier OA_1 will be

$$V_X = -V_{BEQ1}$$

$$V_X = -V_T \ln\left(\frac{I_{CQ1}}{I_S}\right)$$

$$V_X = -V_T \ln\left(\frac{V_1}{R_1 I_S}\right) = -V_T \ln\left(\frac{V_1}{V_R}\right) \tag{2.6}$$

where $R_1 = R$ and $V_R = R I_S$.

Similarly, the output of log amplifier OA_2 will be

$$V_Y = -V_T \ln\left(\frac{V_2}{R_2 I_S}\right) = -V_T \ln\left(\frac{V_2}{V_R}\right) \tag{2.7}$$

The output of adder OA_3 will be

$$V_Z = -(V_X + V_Y) = V_T \ln\left(\frac{V_1 V_2}{V_R^2}\right) \tag{2.8}$$

The antilogarithmic operation of a bipolar transistor is given by

$$I_C = I_S \ln^{-1 V_{BE}/V_T}$$

In Figure 2.2 $V_{BEQ3} = V_Z$ and the collector current of transistor Q_3 will be

$$I_{CQ3} = I_S \ln^{-1 V_Z/V_T}$$

The output voltage is given by

$$V_O = -I_{CQ3} R_4$$

$$V_O = -I_S R_4 \ln^{-1 V_Z/V_T}$$

Let $R_4 = R$, then

$$V_O = -V_R \ln^{-1 V_Z/V_T} \tag{2.9}$$

Equation 2.8 in Equation 2.9 gives

$$V_O = -\frac{V_1 V_2}{V_R} \tag{2.10}$$

2.2 Multiplier Using FETs

An ideal multiplier circuit using FETs as the controlled resistor is shown in Figure 2.3. For small source–drain voltages, an FET acts as a controlled

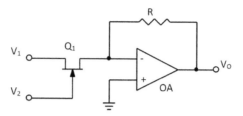

FIGURE 2.3
Basic multiplier using FETs.

resistor whose resistance is inversely proportional to the gate source voltage. The multiplier circuit shown in Figure 2.3 is an amplifier with V_1 as the input voltage whose gain is approximately proportional to V_2. Hence the output voltage is proportional to the product of V_1 and V_2.

The circuit has several major drawbacks including very poor linearity since the gain between V_1 and V_O is not linearly related to V_2. An FET is also very temperature sensitive causing large temperature-dependent errors. There is a restricted range of voltages for V_1 and V_2, i.e., V_1 should be in the range −0.5 V to +0.5 V and V_2 should be 0 V to 0.5 V.

2.3 Variable Transconductance Multiplier

The basic one-quadrant variable transconductance multiplier is shown in Figure 2.4. It uses the principle of the dependence of the transistor trans-conductance on the emitter current bias. The transistors Q_1 and Q_2 form a differential amplifier. For small voltages V_1, i.e., $V_1 \ll V_T$,

$$V_O = g_m R_L V_1$$

where g_m is the transconductance

$$g_m = \frac{I_E}{V_T}$$

In Figure 2.4, if $I_E R_E \gg V_{BE}$, then

$$V_2 = I_E R_E$$

The overall voltage transfer expression can be written as

$$V_O = \frac{I_E}{V_T} R_L V_1 = \frac{V_2}{R_E V_T} R_L V_1$$

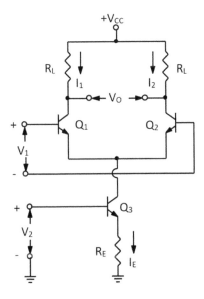

FIGURE 2.4
Circuit diagram of a basic one-quadrant variable transconductance multiplier.

$$V_O = \frac{R_L}{R_E V_T} V_1 V_2$$

$$V_O = \frac{V_1 V_2}{V_R} \qquad (2.11)$$

where $V_R = R_E V_T / R_L$.

2.4 Gilbert Multiplier Cell

The Gilbert cell using an emitter-coupled pair $(Q_1–Q_2)$ in series with a cross-coupled, emitter-coupled pair $(Q_3–Q_6)$ is shown in Figure 2.5. This cell is used to obtain a complete Gilbert cell basic single-quadrant multiplier circuit and is shown in Figure 2.6. V_1 and V_2 are the two input voltages. These two inputs determine the division of the total current I_E among the various branches of the circuit. As the devices are symmetrically cross-coupled and the current I_E is constant, the large common mode shift at the outputs gets eliminated.

Let us assume that all transistors are well matched and h_{fe} for all transistors is very high, i.e., $h_{fe} \gg 1$. From Figure 2.6

$$I_1 + I_2 = I_5 \qquad (2.12)$$

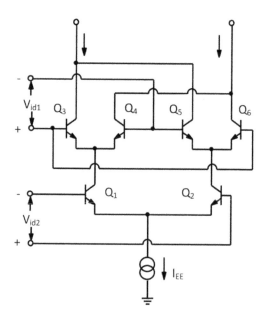

FIGURE 2.5
Circuit diagram of a basic Gilbert cell.

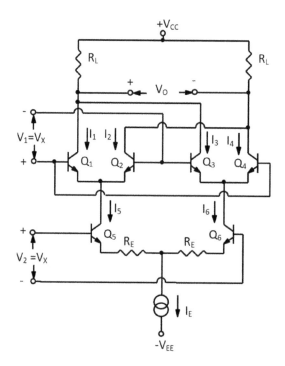

FIGURE 2.6
Basic variable transconductance multiplier using a Gilbert cell.

$$I_3 + I_4 = I_6 \tag{2.13}$$

$$I_5 + I_6 = I_E \tag{2.14}$$

Let us assume $V_1 \ll V_T$, then the current unbalance in the differential pairs can be expressed as

$$I_1 - I_2 = (g_m)_{12} V_1 \tag{2.15}$$

$$I_3 - I_4 = (g_m)_{34} V_1 \tag{2.16}$$

where $(g_m)_{12}$ and $(g_m)_{34}$ are the variable transconductances of the transistor pairs Q_1–Q_2 and Q_3–Q_4, respectively. Under the absence of emitter degeneration resistance, the transconductances are directly proportional to the bias currents I_5 and I_6 and hence

$$(g_m)_{12} = \frac{I_5}{V_T} \tag{2.17}$$

$$(g_m)_{34} = \frac{I_6}{V_T} \tag{2.18}$$

The total differential output voltage V_O is given by

$$V_O = R_L \left[(I_1 - I_2) - (I_3 - I_4) \right] \tag{2.19}$$

Equations 2.15 and 2.16 in Equation 2.19 gives

$$V_O = R_L \left[(g_m)_{12} V_1 - (g_m)_{34} V_1 \right] \tag{2.20}$$

Equations 2.17 and 2.18 in Equation 2.20 gives

$$V_O = \frac{R_L V_1}{V_T} \left[I_5 - I_6 \right] \tag{2.21}$$

If the emitter series resistance R_E is chosen sufficiently high, such that $I_5 R_E \gg V_T$ and $I_6 R_E \gg V_T$, then

$$(I_5 - I_6) = \frac{V_2}{R_E} \tag{2.22}$$

Equation 2.22 in Equation 2.21 gives

$$V_O = \frac{R_L}{R_E V_T} V_1 V_2 \tag{2.23}$$

$$V_O = \frac{V_1 V_2}{V_R} \qquad (2.24)$$

where $V_R = R_E V_T / R_L$.

2.5 Triangle Wave Averaging Multiplier

The block diagram of a triangular wave averaging multiplier is given in Figure 2.7. It is made up of a triangle wave generator, summing amplifiers, diode rectifiers and low pass filters. From the block diagram shown in Figure 2.7,

$$V_X = V_T + V_1 + V_2$$

$$V_Y = V_T + V_1 - V_2$$

The voltage V_X is passed through a positive rectifier to retain the positive portion of the triangular wave. The low pass filter output V_A will be

$$V_A = \frac{1}{T} \int_0^t V_M(t) dt$$

$$V_A = \frac{1}{T} \left[\text{Area of triangle above zero level} \right] = \frac{1}{T} \left[\frac{1}{2} (\text{Base}) (\text{Peak value}) \right]$$

Peak value $= V_1 + V_2 + V_T$

When peak value is V_T, base is $T/2$. Therefore if peak voltage is $V_1 + V_2 + V_T$, then

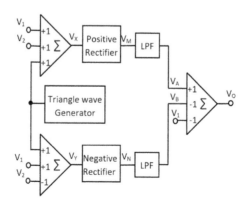

FIGURE 2.7
Block diagram of a triangle averaging multiplier.

$$\text{base} = \frac{(V_1 + V_2 + V_T)\frac{T}{2}}{V_T}$$

$$V_A = \frac{1}{T}\left[\frac{1}{2}\left(\frac{(V_1 + V_2 + V_T)\frac{T}{2}}{V_T}\right)(V_1 + V_2 + V_T)\right]$$

$$V_A = \frac{1}{4V_T}(V_1 + V_2 + V_T)^2 \qquad (2.25)$$

The voltage V_Y is passed through a negative rectifier to retain the negative portion of the triangular wave. The low pass filter output V_B will be

$$V_B = \frac{1}{T}\int_0^t V_N(t)dt$$

$$V_B = \frac{1}{T}\left[\text{Area of triangle below zero level}\right] = \frac{1}{T}\left[\frac{1}{2}(\text{Base})(\text{peakvalue})\right]$$

Only negative voltage of V_Y is passed through. Negative peak value$=V_T-(V_1-V_2)=V_T-V_1+V_2$. When peak value is V_T, base is T/2. Therefore when peak value $V_T-V_1+V_2$, then

$$\text{base} = \frac{(V_T - V_1 + V_2)\frac{T}{2}}{V_T}$$

$$V_B = \frac{1}{T}\left[\frac{1}{2}\left(\frac{(V_T - V_1 + V_2)\frac{T}{2}}{V_T}\right)(V_T - V_1 + V_2)\right]$$

$$V_B = \frac{1}{4V_T}(-V_1 + V_2 + V_T)^2 \qquad (2.26)$$

V_A, V_B and V_1 are combined through a summing amplifier in such a way that

$$V_o = V_A - V_B - V_1 \qquad (2.27)$$

Equations 2.24 and 2.25 in Equation 2.27 gives

$$V_o = \frac{V_1 V_2}{V_T} \qquad (2.28)$$

Figure 2.8 shows the circuit diagram of the proposed triangle averaging multiplier, and its associated waveforms are shown in Figure 2.9.

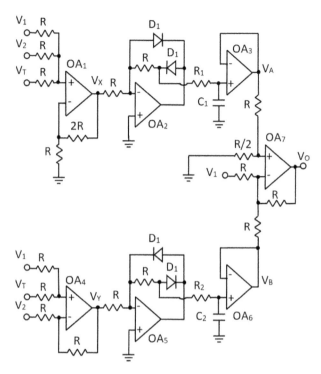

FIGURE 2.8
Circuit diagram of a triangle wave averaging multiplier.

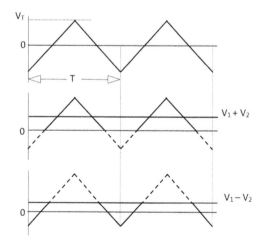

FIGURE 2.9
Associated waveforms of Figure 2.8. From top to bottom: reference triangular wave with peak value V_T, output of positive rectifier V_M and output of negative rectifier V_N.

2.6 Quarter–Square Multiplier

The block diagram of quarter–square multiplication is shown in Figure 2.10. This multiplier is based on the algebraic equation

$$xy = \frac{1}{4}\left[(x+y)^2 - (x-y)^2\right] \qquad (2.29)$$

Let us replace Equation 2.29 with V_1 and V_2 variables

$$V_1 V_2 = \frac{1}{4}\left[(V_1 + V_2)^2 - (V_1 - V_2)^2\right]$$

The adder composed by op-amp OA_1 output will be

$$V_X = V_1 + V_2$$

The subtractor composed by op-amp OA_2 output will be

$$V_Y = V_1 - V_2$$

The output of the positive squarer will be

$$V_A = \left(V_1 + V_2\right)^2$$

The output of the negative squarer will be

$$V_B = \left(V_1 - V_2\right)^2$$

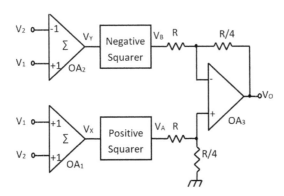

FIGURE 2.10
Block diagram of quarter-square multiplier.

The output of the differential amplifier composed by op-amp OA$_3$ will be

$$V_O = \frac{1}{4}[(V_A) - (V_B)]$$

$$V_O = \frac{1}{4}\left[(V_1 + V_2)^2 - (V_1 - V_2)^2\right]$$

$$V_O = V_1 V_2 \qquad\qquad (2.30)$$

Tutorial Exercises

2.1 Design a multiplier using diode-based log–antilog amplifiers.

2.2 Design a multiplier using FETs.

2.3 Explain the working principles of transistorized variable transconductance multipliers.

2.4 Briefly explain the concept of Gilberts multiplier cells.

2.5 Draw a block diagram of a triangle averaging multiplier and briefly explain.

2.6 Draw a block diagram of a quarter-square multiplier and briefly explain.

2.7 You are given a two squarer (positive and negative squarer). Design a quarter-square multiplier using differential amplifiers.

3

Time Division Multipliers (TDMs): Multiplexing

If the width of a pulse train is made proportional to one voltage and the amplitude of the same pulse train to a second voltage, then the average value of this pulse train is proportional to the product of the two voltages. This is called a time division multiplier or pulse averaging multiplier or sigma delta multiplier. The time division multiplier can be implemented using a (1) triangular wave, (2) saw tooth wave or (3) without using any reference wave.

There are two types of time division multipliers (TDMs): multiplexing TDM (MTDM) and switching TDM (STDM). A time division multiplier using analog 2-to-1 multiplexers is called a multiplexing TDM. A time division multiplier using analog switches is called a switching TDM. Multiplexing time division multipliers are described in this chapter and switching time division multipliers are described in Chapter 5.

3.1 Saw Tooth Wave–Based Time Division Multipliers

Circuit diagrams of saw tooth wave–based time division multipliers are shown in Figure 3.1 and their associated waveforms are shown in Figure 3.2. A saw tooth wave V_{S1} of peak value V_R and time period T is generated by the following:

1. Op-amp OA_1; resistors R_1, R_2 and R_3; capacitor C_1; transistor Q_1; and multiplexer M_1 in Figure 3.1a.

$$V_R = 2V_{BE} \qquad (3.1)$$

$$T = 1.4R_1C_1 \qquad (3.2)$$

2. Op-amps OA_1 and OA_2 and multiplexer M_1 in Figure 3.1b.

In Figure 3.1b, the initial op-amp OA_2 output is zero; the multiplexer M_1 connects 'ax' to 'a'; and the integrator formed by resister R_1, capacitor C_1 and op-amp OA_1 integrates $-V_R$, and its output is given as

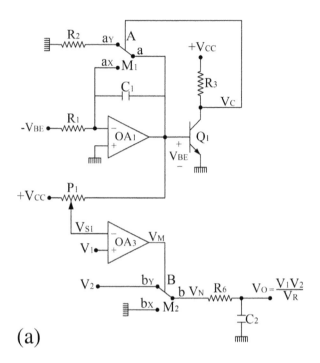

(a)

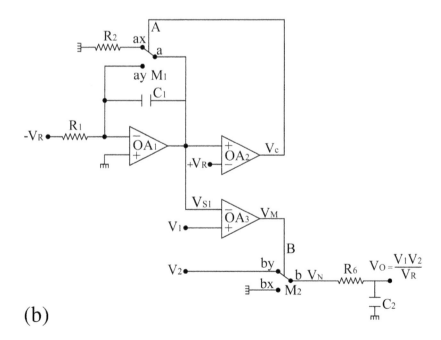

(b)

FIGURE 3.1 (CONTINUED)

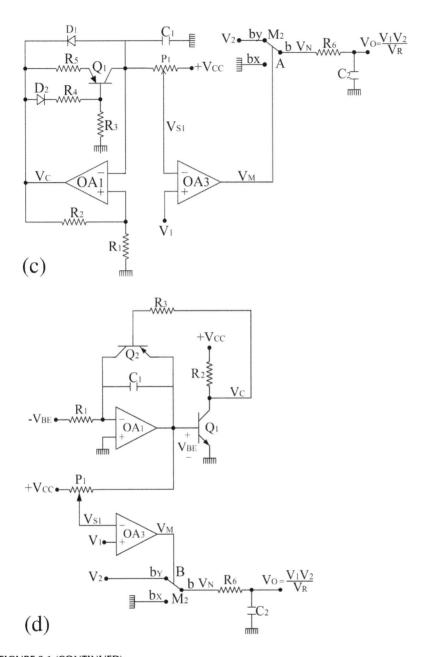

FIGURE 3.1 (CONTINUED)
(a) Saw tooth wave–based time division multiplier type I. (b) Saw tooth wave–based time division multiplier type II. (c) Saw tooth wave–based time division multiplier type III. (d) Saw tooth wave–based time division multiplier type IV.

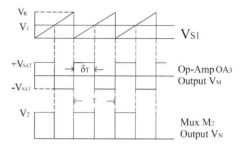

FIGURE 3.2
Associated waveforms of Figure 3.1a–c.

$$V_{S1} = -\frac{1}{R_1C_1}\int -V_R dt$$

$$V_{S1} = \frac{V_R}{R_1C_1}t \qquad (3.3)$$

A positive going ramp is generated at the output of op-amp OA_1 and when it reaches the value of reference voltage V_R, the comparator OA_2 output becomes HIGH. The multiplexer M_1 now connects 'ay' to 'a' and shorts capacitor C_1, and hence the integrator output becomes zero. Then the comparator output is LOW and the sequence therefore repeats to give a perfect saw tooth wave V_{S1} of peak value V_R at op-amp OA_1 output as shown in Figure 3.2. From Equation 3.3, Figure 3.2, and fact that at $t = T$, $V_{S1} = V_R$:

$$V_R = \frac{V_R}{R_1C_1}T$$

Peak value V_R of saw tooth wave = Reference voltage V_R $\qquad (3.4)$

$$T = R_1C_1 \qquad (3.5)$$

3. Op-amp OA_1; transistor Q_1; resistors R_1, R_2, R_3, R_4 and R_5; capacitor C_1; and diodes D_1 and D_2 in Figure 3.1c.
   The peak value V_R is given as

$$V_R = \beta(V_{SAT}) + \frac{\beta(V_{SAT})}{1.5} \qquad (3.6)$$

where β is given as $\beta = \dfrac{R_1}{R_1 + R_2}$.

The time period T of this saw tooth wave V_{S1} is given as

$$T = 2R_5C_1 \ln\left(1 + 2\frac{R_1}{R_2}\right) \quad (3.7)$$

4. Op-amp OA_1; transistors Q_1 and Q_2; resistor R_1; and capacitor C_1 in Figure 3.1d.

$$V_R = 2V_{BE} \quad (3.8)$$

$$T = 2R_1C_1 \quad (3.9)$$

In all circuits in Figure 3.1, the comparator OA_3 compares the saw tooth wave V_{S1} of peak value V_R with the input voltage V_1 and produces a rectangular waveform V_M at its output. The ON time δ_T of this rectangular waveform V_M is given as

$$\delta_T = \frac{V_1}{V_R} T \quad (3.10)$$

The rectangular pulse V_M controls the second multiplexer M_2. When V_M is HIGH, another input voltage V_2 is connected to the R_6C_2 low pass filter ('by' is connected to 'b'). When V_M is LOW, no volts are connected to the R_6C_2 low pass filter ('bx' is connected to 'b'). Another rectangular pulse V_N with maximum value of V_2 is generated at the multiplexer M_2 output. The R_6C_2 low pass filter gives the average value of this pulse train V_N and is given as

$$V_O = \frac{1}{T}\int_0^{\delta_T} V_2 dt$$

$$V_O = \frac{V_2}{T}\delta_T \quad (3.11)$$

Equation 3.10 in Equation 3.11 gives

$$V_O = \frac{V_1 V_2}{V_R} \quad (3.12)$$

Design Exercises

1. Replace multiplexer M_2 in Figure 3.1 with the transistor multiplexer of Figure 1.17, the field-effect transistor (FET) multiplexer of Figure 1.18 and the metal-oxide-semiconductor field-effect transistor (MOSFET) multiplexer of Figure 1.19 (see Chapter 1). In each, (i) draw circuit diagrams, (ii) draw waveforms at appropriate places, (iii) explain the working operation and (iv) deduce expressions for their output voltages.

2. In the multiplier circuits shown in Figure 3.1, reverse the polarity of input voltage V_2. (i) Draw waveforms at appropriate places and (ii) deduce expressions for their output voltages.

3.2 Triangular Wave–Referenced Time Division Multipliers

The circuit diagrams of triangular wave–based multipliers are shown in Figure 3.3 and their associated waveforms are shown in Figure 3.4. In Figure 3.3a a triangular wave V_{T1} with $\pm V_T$ peak-to-peak values and time period T is generated by the op-amp OA_1, and transistors Q_1 and Q_2. The value of V_T is given as

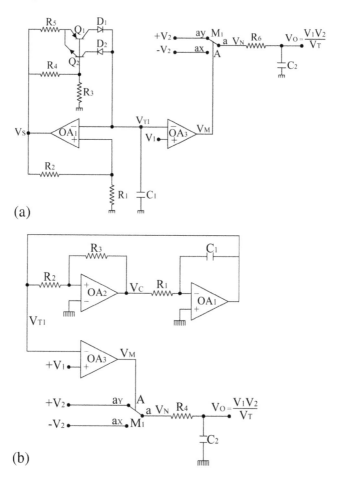

(a)

(b)

FIGURE 3.3
(a) Triangular wave–based multiplier type I. (b) Triangular wave–based multiplier type II.

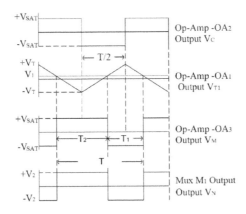

FIGURE 3.4
Associated waveforms of Figure 3.3.

$$V_T = \beta(V_{SAT}) \approx \beta(0.76)(V_{CC}) \tag{3.13}$$

where β is given as $\beta = \dfrac{R_1}{R_1 + R_2}$.

And time period T is given as

$$T = 4R_5C_1 \frac{R_1}{R_2} \tag{3.14}$$

In Figure 3.3b, op-amps OA_1 and OA_2 constitute a triangular/square wave generator. The output of op-amp OA_1 is a triangular wave V_{T1} with $\pm V_T$ peak values and time period of T. If initially the comparator OA_2 output is LOW $(-V_{SAT})$, the output of the integrator composed by op-amp OA_1, resistor R_1 and capacitor C_1 is given as

$$V_{T1} = -\frac{1}{R_1C_1}\int -V_{SAT}dt = \frac{V_{SAT}}{R_1C_1}t \tag{3.15}$$

The integrator output is rising toward positive saturation and when it reaches a value $+V_T$, the comparator output becomes HIGH $(+V_{SAT})$. The output of the integrator composed of op-amp OA_1, resistor R_1 and capacitor C_1 is given as

$$V_{T1} = -\frac{1}{R_1C_1}\int +V_{SAT}dt = -\frac{V_{SAT}}{R_1C_1}t \tag{3.16}$$

Now the output of the integrator is changing its slope from $+V_T$ toward $-V_T$ and when it reaches the value $-V_T$, the comparator output becomes LOW $(-V_{SAT})$ and the sequence therefore repeats to give (1) a triangular waveform V_{T1} with $\pm V_T$ peak-to-peak values at the output of op-amp OA_1 and (2) a square waveform V_C with $\pm V_{SAT}$ peak-to-peak values at the output of comparator OA_2.

From the waveforms shown in Figure 3.4, Equation 3.15, and the fact that at $t = T/2$, $V_{T1} = 2V_T$:

$$2V_T = \frac{V_{SAT}}{R_1 C_1} \frac{T}{2}$$

$$T = \frac{4V_T R_1 C_1}{V_{SAT}} \tag{3.17}$$

When the comparator OA_2 output is LOW ($-V_{SAT}$), the effective voltage at the non-inverting terminal of comparator OA_2 will be by the superposition principle:

$$\frac{(-V_{SAT})}{(R_2 + R_3)} R_2 + \frac{(+V_T)}{(R_2 + R_3)} R_3$$

When this effective voltage at the non-inverting terminal of comparator OA_2 becomes zero

$$\frac{(-V_{SAT})R_2 + (+V_T)R_3}{(R_2 + R_3)} = 0$$

$$(+V_T) = (+V_{SAT}) \frac{R_2}{R_3}$$

When the comparator OA_2 output is HIGH ($+V_{SAT}$), the effective voltage at the non-inverting terminal of comparator OA_2 will be by the superposition principle:

$$\frac{(+V_{SAT})}{(R_2 + R_3)} R_2 + \frac{(-V_T)}{(R_2 + R_3)} R_3$$

When this effective voltage at the non-inverting terminal of comparator OA_2 becomes zero

$$\frac{(+V_{SAT})R_2 + (-V_T)R_3}{(R_2 + R_3)} = 0$$

$$(-V_T) = (-V_{SAT}) \frac{R_2}{R_3}$$

$$\pm V_T = \pm V_{SAT} \frac{R_2}{R_3} \approx 0.76(\pm V_{CC}) \frac{R_2}{R_3} \tag{3.18}$$

From Equations 3.17 and 3.18, time period T of the generated triangular/ square waveform is given by

$$T = 4R_1C_1 \frac{R_2}{R_3} \qquad (3.19)$$

In both circuits of Figure 3.3 one input voltage V_1 is compared with the generated triangular wave V_{T1} by the comparator on OA_3. An asymmetrical rectangular waveform V_M is generated at the comparator OA_3 output. From the waveforms shown in Figure 3.4, it is observed that

$$T_1 = \frac{V_T - V_1}{2V_T} T$$

$$T_2 = \frac{V_T + V_1}{2V_T} T$$

$$T = T_1 + T_2 \qquad (3.20)$$

This rectangular wave V_M is given as the control input to the multiplexer M_1. The multiplexer M_1 connects the other input voltage $+V_2$ during T_2 ('ay' is connected to 'a') and $-V_2$ during T_1 ('ax' is connected to 'a'). Another rectangular asymmetrical wave V_N with a peak-to-peak value of $\pm V_2$ is generated at the multiplexer M_1 output. The R_6C_2 low pass filter gives the average value of the pulse train V_N, which is given as

$$V_O = \frac{1}{T} \left[\int_O^{T_2} V_2 \, dt + \int_{T_2}^{T_1+T_2} (-V_2) \, dt \right] = \frac{V_2}{T} (T_2 - T_1) \qquad (3.21)$$

Equation 3.20 in Equation 3.21 gives

$$V_O = \frac{V_1 V_2}{V_T} \qquad (3.22)$$

Design Exercises

1. The multiplexer M_1 in Figure 3.3 can be replaced with a transistorized multiplexer of Figure 1.17 (see Chapter 1) and shown in Figure 3.5. (i) Explain the working operation of the multipliers shown in Figure 3.5, (ii) draw waveforms at appropriate places and (iii) deduce expressions for the output voltages.

2. The multiplexer M_1 in Figure 3.3 is to be replaced with the FET multiplexers of Figure 1.18 and MOSFET multiplexers of Figure 1.19 (see

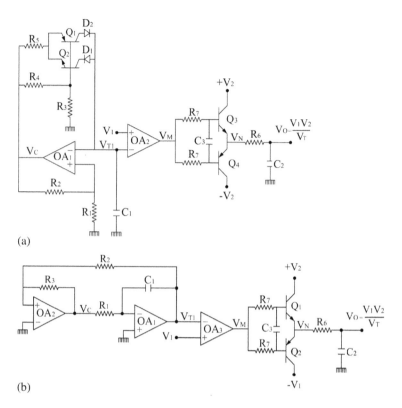

FIGURE 3.5

(a) Triangular wave–based time division multiplier type I. (b) Triangular wave–based time division multiplier type II.

Chapter 1). In each, (i) draw the circuit diagrams, (ii) explain the working operation, (iii) draw waveforms at appropriate places and (iv) deduce expressions for the output voltages.

3.3 Time Division Multiplier with No Reference: Type I

The circuit diagrams of time division multipliers without using either triangular or saw tooth waves as reference are shown in Figure 3.6 and their associated waveforms in Figure 3.7. The op-amps OA_1 and OA_2 along with R_1, C_1, R_2 and R_3 constitute an asymmetrical rectangular wave generator, and a rectangular wave V_C is generated at the output of op-amp OA_2. As discussed in Section 3.2,

$$\pm V_T = \pm V_{SAT}\frac{R_2}{R_3} \simeq 0.76(\pm V_{CC})\frac{R_2}{R_3} \qquad (3.23)$$

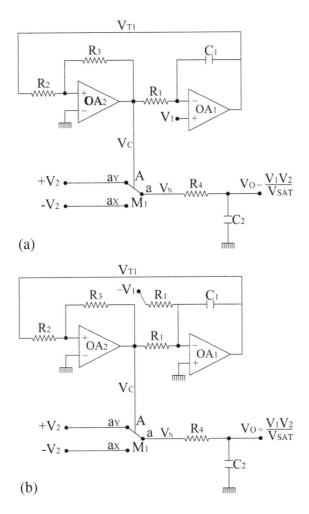

FIGURE 3.6

(a) Time division multiplier with no reference type I. (b) Equivalent circuit of part (a).

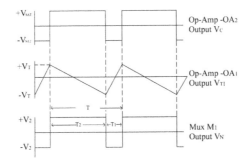

FIGURE 3.7

Associated waveforms of Figure 3.6.

From the waveforms shown in Figure 3.7, it is observed that

$$T_1 = \frac{V_{SAT} - V_1}{2V_{SAT}}$$

$$T_2 = \frac{V_{SAT} + V_1}{2V_{SAT}} T$$

$$T = T_1 + T_2 \tag{3.24}$$

The rectangular wave V_C controls multiplexer M_1, and the multiplexer selects $+V_2$ during T_2 ('ay' is connected to 'a') and $-V_2$ during T_1 ('ax' is connected to 'a') to its output. Another asymmetrical rectangular wave V_N is generated at the multiplexer M_1 output with $\pm V_2$ as the maximum value. The R_4C_2 low pass filter gives the average value of this pulse train V_N and is given as

$$V_O = \frac{1}{T}\left[\int_O^{T_2} V_2\,dt + \int_{T_2}^{T_1+T_2} (-V_2)\,dt \right] = \frac{V_2}{T}[T_2 - T_1] \tag{3.25}$$

Equation 3.24 in Equation 3.25 gives

$$V_O = \frac{V_1 V_2}{V_{SAT}} \tag{3.26}$$

Design Exercises

1. The multiplexer M_1 in Figure 3.6 can be replaced with the transistorized multiplexer of Figure 1.17 (see Chapter 1) and shown in Figure 3.8. (i) Explain the working operation of the multipliers shown in Figure 3.8, (ii) draw waveforms at appropriate places and (iii) deduce expressions for the output voltages.

2. The multiplexer M_1 in Figure 3.6 is to be replaced with the FET multiplexers of Figure 1.18 and MOSFET multiplexers of Figure 1.19 (see Chapter 1). In each, (i) draw the circuit diagram, (ii) explain the working operation, (iii) draw waveforms at appropriate places and (iv) deduce expressions for the output voltages.

3.4 Time Division Multiplier with No Reference: Type II

The circuit diagrams of time division multipliers type II without using either the saw tooth wave or triangular wave as reference are shown in Figure 3.9

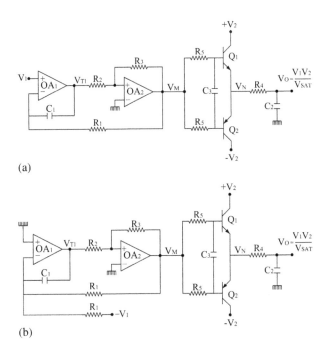

(a)

(b)

FIGURE 3.8
(a) Time division multiplier with no reference type I. (b) Equivalent circuit of part (a).

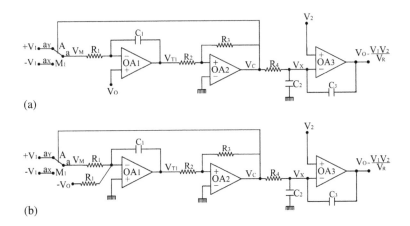

(a)

(b)

FIGURE 3.9
(a) Time division multiplier without reference type II. (b) Equivalent circuit of part (a).

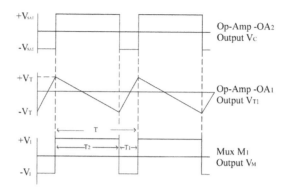

FIGURE 3.10
Associated waveforms of Figure 3.9.

and their associated waveforms are shown in Figure 3.10. The op-amps OA_1 and OA_2 along with R_1, C_1, R_2 and R_3 constitute an asymmetrical rectangular wave generator, and a rectangular wave V_C is generated at the output of op-amp OA_2.

When the comparator OA_2 output is LOW ($-V_{SAT}$), the effective voltage at the non-inverting terminal of comparator OA_2 will be by the superposition principle:

$$\frac{(-V_{SAT})}{(R_2+R_3)}R_2 + \frac{(+V_T)}{(R_2+R_3)}R_3$$

When this effective voltage becomes zero

$$\frac{(-V_{SAT})R_2 + (+V_T)R_3}{(R_2+R_3)} = 0$$

$$(+V_T) = (+V_{SAT})\frac{R_2}{R_3}$$

When the comparator OA_2 output is HIGH ($+V_{SAT}$), the effective voltage at the non-inverting terminal of comparator OA_2 will be by the superposition principle:

$$\frac{(+V_{SAT})}{(R_2+R_3)}R_2 + \frac{(-V_T)}{(R_2+R_3)}R_3$$

When this effective voltage becomes zero

$$\frac{(+V_{SAT})R_2 + (-V_T)R_3}{(R_2+R_3)} = 0$$

$$(-V_T) = (-V_{SAT})\frac{R_2}{R_3}$$

$$\pm V_T = \pm\frac{R_2}{R_3}V_{SAT} \tag{3.27}$$

From the waveforms shown in Figure 3.10. It is observed that

$$T_1 = \frac{V_1 - V_O}{2V_1}T$$

$$T_2 = \frac{V_1 + V_O}{2V_1}T$$

$$T = T_1 + T_2 \tag{3.28}$$

The R_4C_2 low pass filter gives the average value of the rectangular wave V_C and is given as

$$V_X = \frac{1}{T}\left[\int_O^{T_2} V_{SAT}\,dt + \int_{T_2}^{T_1+T_2}(-V_{SAT})\,dt\right] \tag{3.29}$$

$$= \frac{V_{SAT}}{T}[T_2 - T_1]$$

$$V_X = \frac{V_O V_{SAT}}{V_1} \tag{3.30}$$

The op-amp OA_3 is at negative closed feedback configuration and a positive direct current (dc) voltage is ensured in the feedback loop. Hence its non-inverting terminal voltage must equal its inverting terminal voltage:

$$V_2 = V_X \tag{3.31}$$

From Equations 3.27, 3.30 and 3.31

$$V_O = \frac{V_1 V_2}{V_{SAT}}$$

Let us assume $V_R = V_{SAT}$, then

$$V_O = \frac{V_1 V_2}{V_R} \tag{3.32}$$

Design Exercises

1. The multiplexer M_1 in Figure 3.9 can be replaced with a transistorized multiplexer of Figure 1.17 (Chapter 1) and shown in Figure 3.11. (i) Explain the working operation of the multipliers shown in Figure 3.11, (ii) draw waveforms at appropriate places and (iii) deduce expressions for the output voltages.

2. The multiplexer M_1 in Figure 3.9 is to be replaced with the FET multiplexers of Figure 1.18 and MOSFET multiplexers of Figure 1.19 (see Chapter 1). In each, (i) draw the circuit diagrams, (ii) explain the working operation, (iii) draw waveforms at appropriate places and (iv) deduce expressions for the output voltages.

Tutorial Exercises

3.1 For the circuit shown in Figure 3.12, (i) deduce the expression for output voltage V_O, and (ii) draw waveforms of V_{S1}, V_{S2}, V_C, V_M and V_N voltages.

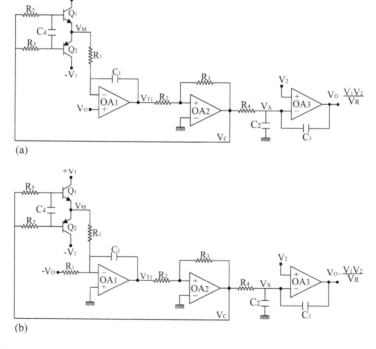

(a)

(b)

FIGURE 3.11
(a) Time division multiplier without reference type II. (b) Equivalent circuit diagram of part (a).

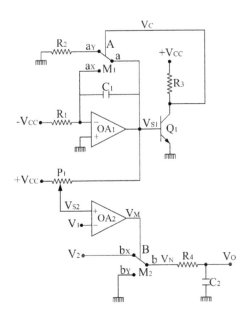

FIGURE 3.12
Circuit for Tutorial Exercise 3.1.

3.2 You are given a saw tooth wave generator block. Deduce the time division multiplier from this saw tooth wave generator.

3.3 You are given a triangular wave generator block. Deduce the time division multiplier from this triangular wave generator.

3.4 In the multiplier circuits shown in Figures 3.3 and 3.6, the inputs ax and ay of multiplexer M_1 are interchanged. (i) Draw waveforms at appropriate places and (ii) deduce expressions of their output voltages.

3.5 In the multiplier circuits shown in Figure 3.9, the polarity of input voltage V_2 is reversed and $-V_O$ is given to op-amp OA_1. (i) Draw waveforms at appropriate places and (ii) deduce expressions of their output voltages.

4

Peak Responding Multipliers: Multiplexing

Peak responding multipliers are classified into peak detecting multipliers and peak sampling multipliers. A short pulse/saw tooth waveform whose time period (T) is proportional to one voltage is generated. Another input voltage is integrated during the time period. The peak value of the integrated voltage is proportional to the product of the two input voltages. This is called a double single-slope peak responding multiplier. A square/triangular waveform whose time period is proportional to one voltage is generated. Another input voltage is integrated during the time period. The peak value of the integrated voltage is proportional to the product of the two input voltages. This is called double dual-slope peak responding multiplier. A rectangular pulse waveform whose OFF time is proportional to one voltage is generated. Another voltage is integrated during this OFF time. The peak value of the integrated output is proportional to the product of the two input voltages. This is called the pulse width integrated peak responding multiplier.

At the output stage of a peak responding multiplier, if a peak detector is used, it is called a peak detecting multiplier, and if sample and hold is used, it is called a peak sampling multiplier. A peak responding multiplier uses either analog switches or analog multiplexers for its operation. If analog switches are used, it is called a switching peak responding multiplier, and if analog multiplexers are used, it is called a multiplexing peak responding multiplier. Multiplexing peak responding multipliers are discussed in this chapter and switching peak responding multipliers in Chapter 6.

4.1 Double Single-Slope Peak Responding Multipliers

The circuit diagrams of peak responding double single-slope multipliers are shown in Figure 4.1 and their associated waveforms are shown in Figure 4.2. Figure 4.1a shows a peak detecting multiplier and Figure 4.1b shows a peak sampling multiplier. When the op-amp OA_2 output is LOW, the multiplexer M_1 connects 'ax' to 'a', and an integrator formed by resistor R_1, capacitor C_1 and op-amp OA_1 integrates the reference voltage $-V_R$. The integrated output will be

$$V_{S1} = -\frac{1}{R_1 C_1} \int -V_R dt = \frac{V_R}{R_1 C_1} t \qquad (4.1)$$

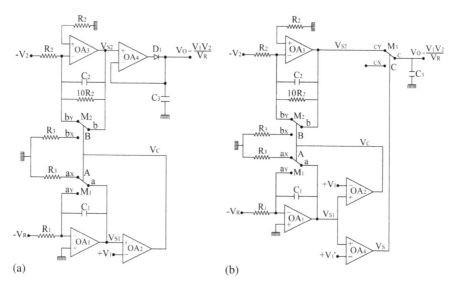

FIGURE 4.1
(a) Double single-slope peak detecting multiplier. (b) Double single-slope peak sampling multiplier.

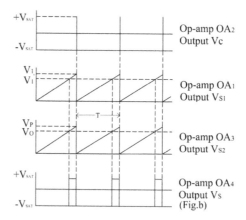

FIGURE 4.2
Associated waveforms of Figure 4.1.

A positive going ramp V_{s_1} is generated at the output of op-amp OA_1. When the output of OA_1 reaches the voltage level of V_1, the comparator OA_2 output becomes HIGH. The multiplexer M_1 connects 'ay' to 'a' and hence the capacitor C_1 is shorted so that op-amp OA_1 output becomes ZERO. Then op-amp OA_2 output goes to LOW, the multiplexer M_1 connects 'ax' to 'a', and the integrator composed by R_1, C_1 and op-amp OA_1 integrates the reference voltage $-V_R$, and the cycle therefore repeats to provide (1) a saw tooth wave of peak value V_1 at the output of op-amp OA_1 and (2) a short pulse waveform V_C at the output of comparator OA_2. The short pulse V_C also controls multiplexer M_2. During the short HIGH time of V_C, multiplexer M_2 selects 'by' to 'b', and the capacitor C_2 is short-circuited so that op-amp OA_3 output is zero volts. During LOW time of V_C, multiplexer M_2 selects 'bx' to 'b', and the integrator formed by resistor R_2, capacitor C_2 and op-amp OA_3 integrates its input voltage $-V_2$, and its output is given as

$$V_{S2} = -\frac{1}{R_2C_2}\int -V_2 dt = \frac{V_2}{R_2C_2}t \tag{4.2}$$

Another saw tooth waveform V_{S2} with peak value V_P is generated at the output of op-amp OA_3. From the waveforms shown in Figure 4.2, from Equations 4.1 and 4.2, and the fact that at $t=T$, $V_{S1}=V_1$, $V_{S2}=V_P$:

$$V_1 = \frac{V_R}{R_1C_1}T \tag{4.3}$$

$$V_P = \frac{V_2}{R_2C_2}T \tag{4.4}$$

From Equations 4.3 and 4.4

$$V_P = \frac{V_2}{R_2C_2}\frac{V_1}{V_R}R_1C_1$$

Let us assume $R_1=R_2$ and $C_1=C_2$, then

$$V_P = \frac{V_1V_2}{V_R} \tag{4.5}$$

(i) In the circuit shown in Figure 4.1a, the peak detector realized by op-amp OA_4, diode D_1 and capacitor C_3 gives the peak value V_P at its output V_O: $V_O=V_P$.

(ii) In the circuit shown in Figure 4.1b, the peak value V_P is obtained by the sample and hold circuit realized by multiplexer M_3 and capacitor C_3. The sampling pulse is generated by op-amp OA_4 by comparing a slightly less than voltage of V_1, called V_1' with the saw tooth wave V_{S1}. The sample and hold operation is illustrated graphically in Figure 4.2. The sample and hold output is $V_O=V_P$.

From Equation 4.5, the output voltage is given as $V_O=V_P$:

$$V_O = \frac{V_1V_2}{V_R} \tag{4.6}$$

Design Exercises

1. In the multiplier circuits shown in Figure 4.1, if the polarity of input voltage V_2 and direction of diode D_1 are reversed, (i) draw waveforms at appropriate places and (ii) derive expressions for their output voltages.

2. In the multiplier circuits shown in Figures 4.1, if the polarity of V_R and V_1 are reversed, (i) draw waveforms at appropriate places and (ii) derive expressions for their output voltages.

4.2 Double Dual-Slope Peak Responding Multipliers: Type I

The circuit diagrams of double dual-slope peak responding multipliers type I are shown in Figure 4.3 and their associated waveforms are in Figure 4.4. Figure 4.3a shows a peak detecting multiplier and Figure 4.3b shows a peak sampling multiplier. When comparator OA_2 output is LOW, $-V_1$ is given to the integrator formed by resistor R_1, capacitor C_1 and op-amp OA_1 by the multiplexer M_1 ('ax' is connected to 'a'); and $-V_O$ is given to the integrator formed by resistor R_4, capacitor C_2 and op-amp OA_3 by the multiplexer M_2 ('bx' is connected to 'b'). The output of op-amp OA_1 will be

$$V_{T1} = -\frac{1}{R_1 C_1} \int -V_1 dt = \frac{V_1}{R_1 C_1} t \tag{4.7}$$

The output of op-amp OA_3 will be

$$V_{T2} = -\frac{1}{R_4 C_2} \int -V_O dt = \frac{V_O}{R_4 C_2} t \tag{4.8}$$

The output of op-amp OA_1 is a positive going ramp, and when it reaches a value $+V_T$ the comparator OA_2 output becomes HIGH. The multiplexer M_1 selects $+V_1$ to op-amp integrator OA_1 ('ay' is connected to 'a') and the multiplexer M_2 selects $+V_O$ to the op-amp integrator OA_3 ('by' is connected to 'b').
The output of op-amp OA_1 will now be

$$V_{T1} = -\frac{1}{R_1 C_1} \int V_1 dt = -\frac{V_1}{R_1 C_1} t$$

The output of op-amp OA_3 will be

$$V_{T2} = -\frac{1}{R_4 C_2} \int V_O dt = -\frac{V_O}{R_4 C_2} t$$

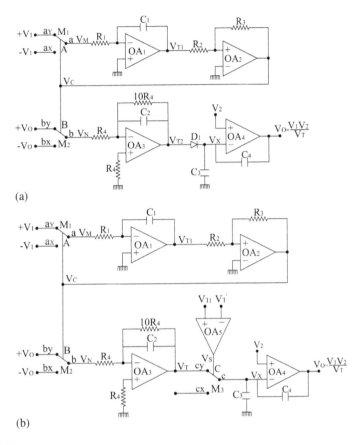

(a)

(b)

FIGURE 4.3
(a) Double dual-slope peak detecting multiplier type I. (b) Double dual-slope peak sampling multiplier type I.

The output of op-amp OA_1 changes slope from $+V_T$ toward $-V_T$ and when it reaches a value of $-V_T$, the comparator OA_2 output becomes LOW and the cycle therefore repeats to give (1) a triangular wave V_{T1} at the output of op-amp OA_1 with $\pm V_T$ peak-to-peak values, (2) another triangular wave V_{T2} at the output of op-amp OA_3 with $\pm V_P$ peak-to-peak values, (3) first square waveform V_C with $\pm V_{SAT}$ peak-to-peak values at the output of comparator OA_2, (4) second square waveform V_M with $\pm V_1$ as peak-to-peak values at the output of multiplexer M_1 and (5) third square waveform V_N with $\pm V_O$ as peak-to-peak values at the output of multiplexer M_2.

From Equations 4.7 and 4.8, the waveforms shown in Figure 4.4, and the fact that at $t = T/2$, $V_{T1} = 2V_T$, $V_{T2} = 2V_P$:

$$2V_T = \frac{V_1}{R_1 C_1} \frac{T}{2} \tag{4.9}$$

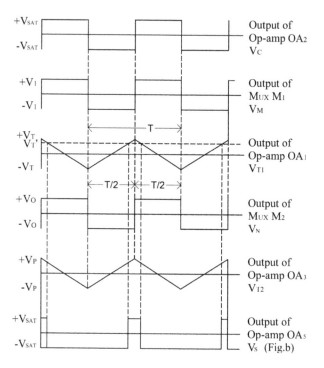

FIGURE 4.4
Associated waveforms of Figure 4.3.

$$2V_P = \frac{V_O}{R_4C_2} \frac{T}{2} \qquad (4.10)$$

From Equations 4.9 and 4.10

$$2V_P = \frac{V_O}{R_4C_2} \frac{2V_TR_1C_1}{V_1}$$

Let us assume $R_1 = R_4$ and $C_1 = C_2$, then

$$V_P = \frac{V_O}{V_1} V_T \qquad (4.11)$$

When the comparator OA_2 output is LOW ($-V_{SAT}$), the effective voltage at the non-inverting terminal of comparator OA_2 will be by the superposition principle:

$$\frac{(-V_{SAT})}{(R_2 + R_3)} R_2 + \frac{(+V_T)}{(R_2 + R_3)} R_3$$

When this effective voltage at the non-inverting terminal of comparator OA_2 becomes zero

$$\frac{(-V_{SAT})R_2 + (+V_T)R_3}{(R_2 + R_3)} = 0$$

$$(+V_T) = (+V_{SAT})\frac{R_2}{R_3}$$

When the comparator OA_2 output is HIGH ($+V_{CC}$), the effective voltage at the non-inverting terminal of comparator OA_2 will be by the superposition principle:

$$\frac{(+V_{SAT})}{(R_2 + R_3)}R_2 + \frac{(-V_T)}{(R_2 + R_3)}R_3$$

When this effective voltage at the non-inverting terminal of comparator OA_2 becomes zero

$$\frac{(+V_{SAT})R_2 + (-V_T)R_3}{(R_2 + R_3)} = 0$$

$$(-V_T) = (-V_{SAT})\frac{R_2}{R_3}$$

$$\pm V_T = \pm\frac{R_2}{R_3}V_{SAT} \simeq \pm\frac{R_2}{R_3}0.76V_{CC} \qquad (4.12)$$

1. In the circuit shown in Figure 4.3a, the peak detector D_1 and capacitor C_3 give the peak value V_P of the triangular wave V_{T2}, i.e., $V_X = V_P$.
2. In the circuit shown in Figure 4.3b, the peak value V_P is obtained by the sample and hold circuit realized by multiplexer M_3 and capacitor C_3. The sampling pulse V_S is generated by op-amp OA_5 by comparing a slightly less than voltage of V_T called V_T' with the triangular wave V_{T1}. The sample and hold operation is illustrated graphically in Figure 4.4. The sampled output is given as $V_X = V_P$.

The op-amp OA_4 is kept in a negative closed loop configuration and a positive direct current (dc) voltage is ensured in the feedback. Hence its inverting terminal voltage will equal its non-inverting terminal voltage, i.e.,

$$V_X = V_2 = V_P \qquad (4.13)$$

From Equations 4.11 and 4.13

$$V_O = \frac{V_1 V_2}{V_T} \qquad (4.14)$$

Design Exercises

1. The multiplexers M_1 and M_2 in Figure 4.3 can be replaced with the transistorized multiplexers of Figure 1.17 (see Chapter 1) and shown in Figure 4.5. (i) Explain the working operation of the multipliers shown in Figures 4.5, (ii) draw waveforms at appropriate places and (iii) deduce expressions for their outputs.

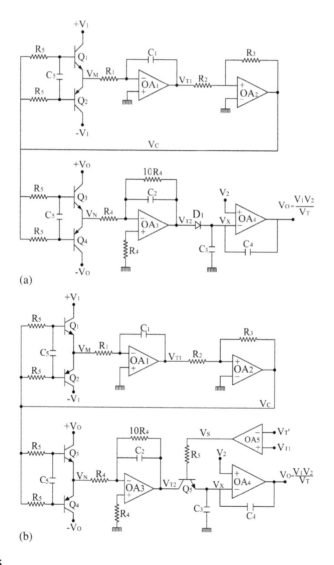

FIGURE 4.5
(a) Double dual-slope peak detecting multiplier type I. (b) Double dual-slope peak sampling multiplier type I.

2. The multiplexers M_1 and M_2 in Figure 4.3 are to be replaced with the FET multiplexers of Figure 1.18 and MOSFET multiplexers Figure 1.19 (see Chapter 1). In each, (i) draw the circuit diagrams, (ii) explain their working operation, (iii) draw waveforms at appropriate places and (iv) deduce the expression for the output voltage.

4.3 Double Dual-Slope Peak Responding Multipliers: Type II

The circuit diagrams of double dual-slope peak responding multipliers type II are shown in Figure 4.6 and their associated waveforms are shown in Figure 4.7. Figure 4.6a shows a double dual-slope peak detecting multiplier and Figure 4.6b shows a double dual-slope peak sampling multiplier. Let's assume the comparator OA_2 output is LOW. The multiplexer M_1 selects $-V_1$ to one end of resistor R_3 ('ax' is connected to 'a'), and $-V_{SAT}$ is given to the integrator formed by resistor R_1, capacitor C_1 and op-amp OA_1. The multiplexer M_2 selects $-V_2$ to the integrator formed by resistor R_4, capacitor C_2 and op-amp OA_3 ('bx' is connected to 'b'). The op-amp OA_1 output is given as

$$V_{T1} = -\frac{1}{R_1C_1}\int -V_{SAT}\,dt = \frac{V_{SAT}}{R_1C_1}t \qquad (4.15)$$

The op-amp OA_3 output is given as

$$V_{T2} = -\frac{1}{R_4C_2}\int -V_2\,dt = \frac{V_2}{R_4C_2}t \qquad (4.16)$$

The output of op-amp OA_1 is rising toward positive saturation and when it reaches a value of $+V_T$, the comparator OA_2 output becomes HIGH. The multiplexer M_1 selects $+V_1$ to one end of resistor R_3 ('ay' is connected to 'a'), and $+V_{SAT}$ is given to the integrator formed by R_1, C_1 and op-amp OA_1. The multiplexer M_2 selects $+V_2$ to the integrator formed by R_4, C_2 and op-amp OA_3 ('by' is connected to 'b'). The output of OA_1 will now be

$$V_{T1} = -\frac{1}{R_1C_1}\int V_{SAT}\,dt = -\frac{V_{SAT}}{R_1C_1}t \qquad (4.17)$$

And the output of OA_3 will be

$$V_{T2} = -\frac{1}{R_4C_2}\int V_2\,dt = -\frac{V_2}{R_4C_2}t \qquad (4.18)$$

The output of op-amp OA_1 changes its slope and moves toward negative saturation. When the output of op-amp integrator OA_1 comes down to a value $-V_T$, the comparator OA_2 output will become LOW and therefore the cycle

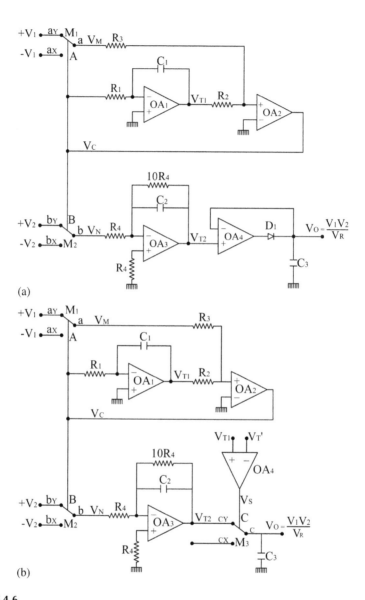

(a)

(b)

FIGURE 4.6
(a) Double dual-slope peak detecting multiplier type II. (b) Double dual-slope peak sampling multiplier type II.

repeats itself to give (1) a triangular wave V_{T1} at the output of op-amp OA_1 with $\pm V_T$ peak-to-peak values, (2) another triangular wave V_{T2} at the output of op-amp OA_3 with $\pm V_P$ peak-to-peak values, (3) first square waveform V_C with $\pm V_{SAT}$ peak-to-peak values at the output of op-amp OA_2, (4) second square waveform V_M with $\pm V_1$ as peak-to-peak values at the output of multiplexer M_1 and (5) third square waveform V_N with $\pm V_2$ as peak-to-peak values

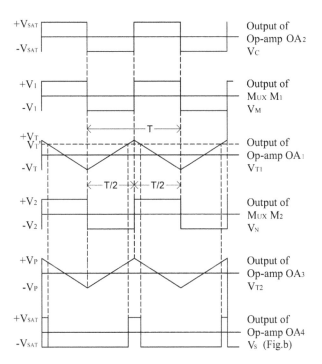

FIGURE 4.7
Associated waveforms of Figure 4.6.

at the output of multiplexer M_2. From the waveforms shown in Figure 4.7, from Equations 4.15 and 4.16, and the fact that at $t = T/2$, $V_{T1} = 2V_T$, $V_{T2} = 2V_P$:

$$2V_T = \frac{V_{SAT}}{R_1 C_1} \frac{T}{2} \qquad (4.19)$$

$$2V_P = \frac{V_2}{R_4 C_2} \frac{T}{2} \qquad (4.20)$$

From Equations 4.19 and 4.20

$$2V_P = \frac{V_2}{R_4 C_2} \frac{2V_T R_1 C_1}{V_{SAT}}$$

Let us assume $R_1 = R_4$ and $C_1 = C_2$, then

$$V_P = \frac{V_2}{V_{SAT}} V_T \qquad (4.21)$$

When the comparator OA_2 output is LOW ($-V_{SAT}$), $-V_1$ will be at the output of multiplexer M_1, and the effective voltage at the non-inverting terminal of comparator OA_2 will be by the superposition principle:

$$\frac{(-V_1)}{(R_2+R_3)}R_2 + \frac{(+V_T)}{(R_2+R_3)}R_3$$

When this effective voltage at the non-inverting terminal of comparator OA_2 becomes zero

$$\frac{(-V_1)R_2+(+V_T)R_3}{(R_2+R_3)} = 0$$

$$(+V_T) = (+V_1)\frac{R_2}{R_3}$$

When the comparator OA_2 output is HIGH $(+V_{CC})$, the effective voltage at the non-inverting terminal of comparator OA_2 will be by the superposition principle

$$\frac{(+V_1)}{(R_2+R_3)}R_2 + \frac{(-V_T)}{(R_2+R_3)}R_3$$

When this effective voltage at the non-inverting terminal of comparator OA_2 becomes zero

$$\frac{(+V_1)R_2+(-V_T)R_3}{(R_2+R_3)} = 0$$

$$(-V_T) = (-V_1)\frac{R_2}{R_3}$$

$$\pm V_T = \pm V_1\frac{R_2}{R_3} \qquad (4.22)$$

Equation 4.22 in Equation 4.21 gives

$$V_P = \frac{V_1 V_2}{V_{SAT}}\frac{R_2}{R_3}$$

Let

$$V_R = \frac{V_{SAT}}{R_2}R_3$$

$$V_P = \frac{V_1 V_2}{V_R} \qquad (4.23)$$

1. In the circuit shown in Figure 4.6a, the peak detector realized by op-amp OA_4, diode D_1 and capacitor C_3 gives the peak value V_P at the output: $V_O = V_P$.

2. In the circuit shown in Figure 4.6b, the peak value V_P is obtained by the sample and hold circuit realized by multiplexer M_3 and capacitor C_3. The sampling pulse V_S is generated by op-amp OA_4 by comparing a slightly less than voltage of V_T called V_T' with the triangular wave V_{T1}. The sample and hold operation is illustrated graphically in Figure 4.7. The sampled output is given as $V_O = V_P$.

From Equation 4.23, the output voltage is given as $V_O = V_P$:

$$V_O = \frac{V_1 V_2}{V_R} \qquad (4.24)$$

Design Exercises

1. The multiplexers M_1 and M_2 in Figure 4.6 can be replaced with the transistorized multiplexer of Figure 1.17 (see Chapter 1) and shown in Figure 4.8. (i) Explain the working operation of the multipliers shown in Figure 4.8, (ii) draw waveforms at appropriate places and (iii) deduce expressions for their outputs.
2. The multiplexers M_1 and M_2 in Figure 4.6 are to be replaced with the FET multiplexers of Figure 1.18 and the MOSFET multiplexers of Figure 1.19 (see Chapter 1). In each, (i) draw the circuit diagrams, (ii) explain their working operation, (iii) draw waveforms at appropriate places and (iv) deduce the expression for the output voltage.

4.4 Pulse Width Integrated Peak Responding Multipliers

The circuit diagrams of pulse width integrated peak responding multipliers are shown in Figure 4.9 and their associated waveforms are shown in Figure 4.10. Figure 4.9a shows peak detecting multiplier type I, and Figure 4.9b shows peak detecting multiplier type II, Figure 4.9c shows peak sampling multiplier type I and Figure 4.9d shows peak sampling multiplier type II.

In Figure 4.9a and c, as discussed in Chapter 3 (Section 3.1, Figure 3.1b), the op-amps OA_1 and OA_2 and multiplexer M_1 constitute a saw tooth wave generator, and a saw tooth wave V_{S1} of peak value V_R is generated at the output of OA_1. The time period of this saw tooth wave is given as from Equation 3.4:

$$T = R_1 C_1 \qquad (4.25)$$

Peak value = Reference voltage V_R

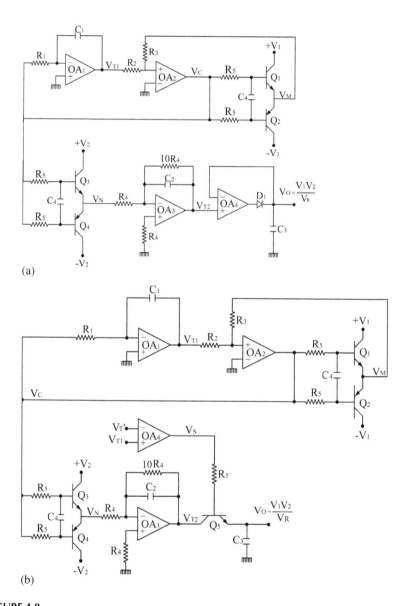

FIGURE 4.8
(a) Double dual-slope peak detecting multiplier type II. (b) Double dual-slope peak sampling multiplier type II.

In Figure 4.9b and d, as discussed in Chapter 3 (Section 3.1, Figure 3.1a), the op-amp OA_1, transistor Q_1 and multiplexer M_1 constitutes saw tooth wave generator and a saw tooth wave V_{S1} of peak value V_R is generated at the output of op-amp OA_1. The time period of this saw tooth wave is given as from Equation 3.2:

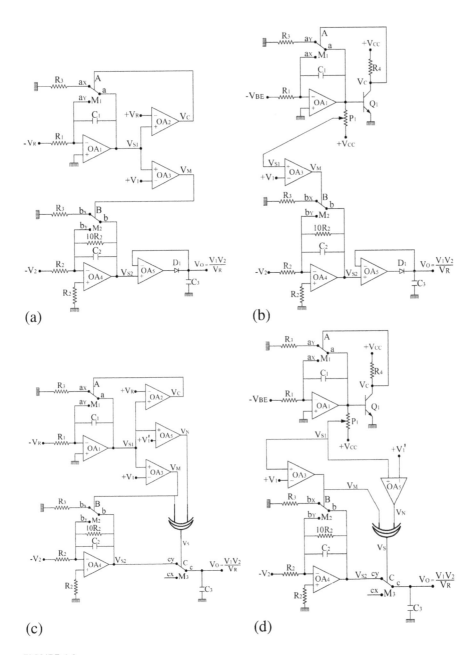

FIGURE 4.9
(a) Pulse width integrated peak detecting multiplier type I. (b) Pulse width integrated peak detecting multiplier type II. (c) Pulse width integrated peak sampling multiplier type 1. (d) Pulse width integrated peak sampling multiplier type II.

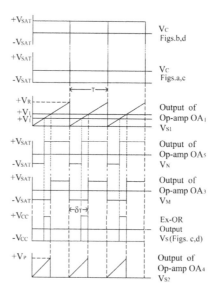

FIGURE 4.10
Associated waveforms of Figure 4.9.

$$T = 1.4R_1C_1 \tag{4.26}$$

$$V_R = 2V_{BE}$$

In Figure 4.9, the saw tooth waveform V_{S1} is compared with one input voltage V_1 by the comparator OA_3. An asymmetrical rectangular wave V_M is generated at the output of comparator OA_3. The OFF time of this wave V_M is given as

$$\delta_T = \frac{V_1}{V_R}T \tag{4.27}$$

The output of comparator OA_3 is given as the control input of multiplexer M_2. During the ON time of V_M, the multiplexer M_2 selects 'by' to 'b' and the capacitor C_2 is shorted so that zero volts appear at op-amp OA_4 output. During the OFF time of V_M, the multiplexer M_2 selects 'bx' to 'b', and another integrator is formed by resistor R_2, capacitor C_2 and op-amp OA_4. This integrator integrates the input voltage $(-V_2)$ and its output is given as

$$V_{S2} = -\frac{1}{R_2C_2}\int -V_2\,dt = \frac{V_2}{R_2C_2}t \tag{4.28}$$

A semi saw tooth wave V_{S2} with peak value of V_P is generated at the output of op-amp OA_4. From the waveforms shown in Figure 4.10, Equation 4.26, and fact that at $t = \delta_T$, $V_{S2} = V_P$:

$$V_P = \frac{V_2}{R_2C_2}\delta_T \tag{4.29}$$

Equations 4.26 and 4.27 in Equation 4.29 gives

$$V_P = \frac{V_1V_2}{V_R}\frac{R_1C_1}{R_2C_2} \tag{4.30}$$

Let $R_1 = R_2$ and $C_1 = C_2$, then

$$V_P = \frac{V_1V_2}{V_R} \tag{4.31}$$

1. In the circuits shown in Figure 4.9a and b the peak detector realized by diode D_1 and capacitor C_3 gives the peak value V_P at its output: $V_O = V_P$.
2. In the circuits shown in Figure 4.9c and d the peak value V_P is obtained by the sample and hold circuit realized by multiplexer M_3 and capacitor C_3. The sampling pulse V_S is generated by the Ex-OR gate from the signals V_M and V_N. V_N is obtained by comparing the slightly less than voltage of V_1, i.e., V_1', with the saw tooth waveform V_{S1}. The sample and hold operation is illustrated graphically in Figure 4.10. The sampled output is given as $V_O = V_P$.

From Equation 4.31, the output voltage is given as $V_O = V_P$:

$$V_O = \frac{V_1V_2}{V_R} \tag{4.32}$$

Design Exercise

Replace the saw tooth generators in Figure 4.9 with the saw tooth wave generators shown in Figure 3.1c and d (see Chapter 3). In each, (i) draw the circuit diagrams, (ii) explain their working operation, (iii) draw waveforms at appropriate places and (iv) deduce the expression for the output voltage.

Tutorial Exercises

4.1 In the multiplier circuit shown in Figure 4.1b, the polarity of V_R and V_1 are to be opposite. Justify it.

4.2 In the multiplier circuits shown in Figure 4.6, if the inputs of multiplexer M_2 are interchanged and the direction of diode D_1 is reversed, (i) draw waveforms at appropriate places and (ii) deduce expressions for their output voltages.

4.3 In the multiplier circuits shown in Figure 4.9a and b, if the polarity of V_2 and direction of diode D_1 are reversed, (i) draw waveforms at appropriate places and (ii) deduce expressions for their output voltages.

5

Time Division Multipliers (TDMs): Switching

As discussed in Chapter 3, if the width of a pulse train is made proportional to one voltage and the amplitude of the same pulse train to a second voltage, then the average value of this pulse train is proportional to the product of the two voltages and is called a time division multiplier or pulse averaging multiplier or sigma delta multiplier. The time division multiplier can be implemented using a triangular wave, a saw tooth wave or without using any reference clock.

There are two types of time division multipliers (TDMs): multiplexing TDMs (MTDMs) and switching TDMs (STDMs). A time division multiplier using analog 2-to-1 multiplexers is called a multiplexing TDM. A time division multiplier using analog switches is called a switching TDM. Multiplexing time division multipliers are described in Chapter 3 and switching time division multipliers are described in this chapter.

5.1 Saw Tooth Wave–Based Time Division Multipliers: Type I

The circuit diagrams of saw tooth wave–based time division multipliers type I are shown in Figure 5.1 and their associated waveforms in Figure 5.2. Figure 5.1a shows a series switching TDM and Figure 5.1b shows a shunt or parallel switching TDM. As discussed in Chapter 3 (Section 3.1, Figure 3.1c), the op-amp OA_1 along with transistor Q_1; resistors R_1, R_2, R_3, R_4 and R_5; diodes D_1 and D_2; and capacitor C_1 constitute a saw tooth wave generator, and a saw tooth wave V_{S1} with peak value V_R and time period T is generated at its output.

The time period T of this saw tooth wave V_{S1} is given as

$$T = 2R_5C_1 \ln\left(1+2\frac{R_1}{R_2}\right)$$

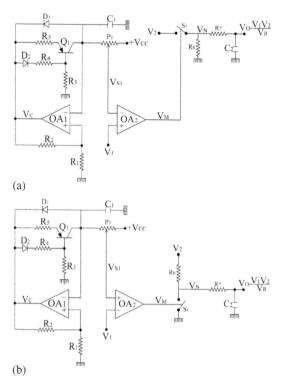

(a)

(b)

FIGURE 5.1
(a) Series switching saw tooth wave–based time division multiplier type I. (b) Shunt switching saw tooth wave–based time division multiplier type I.

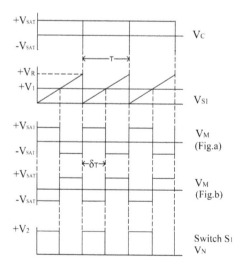

FIGURE 5.2
Associated waveforms of Figure 5.1.

The peak value V_R is given as

$$V_R = \beta(V_{SAT}) + \frac{\beta(V_{SAT})}{1.5} \tag{5.1}$$

where β is given as $\beta = \dfrac{R_1}{R_1 + R_2}$.

The comparator OA_2 compares this saw tooth wave V_{S1} with an input voltage V_1 and produces a rectangular waveform V_M.

The ON time (Figure 5.1a) or the OFF time (Figure 5.1b) of V_M is given as

$$\delta_T = \frac{V_1}{V_R} T \tag{5.2}$$

The rectangular pulse V_M controls the switch S_1.

1. In Figure 5.1a, when V_M is HIGH, the switch S_1 is closed and another input voltage V_2 is connected to the R_7C_2 low pass filter. When V_M is LOW, the switch S_1 is opened and zero volts exist on the R_7C_2 low pass filter.

2. In Figure 5.1b, when V_M is HIGH, the switch S_1 is closed and zero volts exist on the R_7C_2 low pass filter. When V_M is LOW, the switch S_1 is opened and another input voltage V_2 is connected to the R_7C_2 low pass filter.

Another rectangular pulse V_N with maximum value of V_2 is generated at the switch S_1 output. The R_7C_2 low pass filter gives the average value of this pulse train V_N and is given as

$$V_O = \frac{1}{T} \int_0^{\delta_T} V_2 \, dt = \frac{V_2}{T} \delta_T \tag{5.3}$$

Equation 5.2 in Equation 5.3 gives

$$V_O = \frac{V_1 V_2}{V_R} \tag{5.4}$$

Design Exercises

1. The switch S_1 in Figure 5.1 is replaced with the transistor switches of Figure 1.13 (see Chapter 1) and shown in Figure 5.3. (i) Explain working operation of the multipliers shown in Figure 5.3, (ii) draw waveforms in appropriate places and (iii) deduce expressions for their outputs.

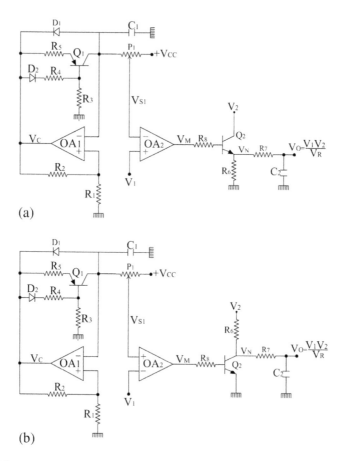

FIGURE 5.3

(a) Series switch saw tooth wave–based time division multiplier type I. (b) Shunt switch saw tooth–based time division multiplier type I.

2. The switch S_1 in Figure 5.1 is to be replaced with the FET switches shown in Figure 1.14 and MOSFET switches shown in Figure 1.15 (see Chapter 1). In each, (i) draw the circuit diagrams, (ii) explain their working operations, (iii) draw waveforms at appropriate places and (iv) deduce expressions for their output voltages.

3. The saw tooth generator in part of Figure 5.1 is to be replaced with the saw tooth generator of Figure 3.1d (see Chapter 3). (i) Draw the circuit diagrams, (ii) explain their working operations, (iii) draw waveforms at appropriate places and (iv) deduce expressions for their output voltages.

4. The switch S_1 in the designed multiplier of Exercise 3 is to be replaced with FET switches shown in Figure 1.14 and MOSFET switches

shown in Figure 1.15 (see Chapter 1). In each, (i) draw the circuit diagrams, (ii) explain their working operations, (iii) draw waveforms at appropriate places and (iv) deduce expressions for their output voltages.

5.2 Saw Tooth Wave–Based Time Division Multipliers: Type II

The circuit diagrams of a saw tooth wave–based multiplier type II are shown in Figure 5.4 and their associated waveforms are shown in Figure 5.5.

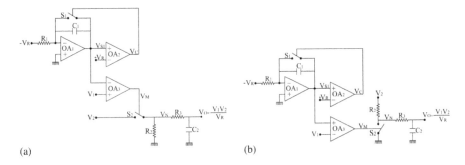

(a) (b)

FIGURE 5.4
(a) Series switching saw tooth wave–based time division multiplier type II. (b) Shunt switching saw tooth wave–based time division multiplier type II.

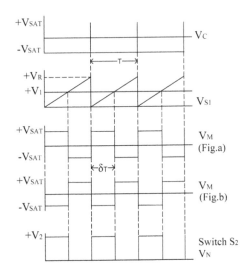

FIGURE 5.5
Associated waveforms of Figure 5.4.

Figure 5.4a shows a series switching TDM and Figure 5.4b shows a shunt switching TDM. When the op-amp OA_2 output is LOW, the switch S_1 is opened and the integrator formed by resister R_1, capacitor C_1 and op-amp OA_1 integrates the reference voltage $-V_R$ and its output is given as

$$V_{S1} = -\frac{1}{R_1C_1}\int -V_R\, dt$$

$$V_{S1} = \frac{V_R}{R_1C_1}t \qquad\qquad (5.5)$$

A positive going ramp is generated at the output of op-amp OA_1 and when it reaches the value of reference voltage $+V_R$ the comparator OA_2 output becomes HIGH. The switch S_1 is closed and shorts capacitor C_1 and hence the integrator output becomes zero. Then the comparator output is LOW, and the sequence therefore repeats to give a perfect saw tooth wave V_{S1} of peak value V_R at op-amp OA_1 output and a short pulse waveform V_C at op-amp OA_2 output as shown in Figure 5.4. From Equation 5.5, Figure 5.5, and the fact that at $t = T$, $V_{S1} = +V_R$:

$$V_R = \frac{V_R}{R_1C_1}T$$

$$T = R_1C_1 \qquad\qquad (5.6)$$

The comparator OA_3 compares one input voltage V_1 with the saw tooth wave V_{S1} and produces a rectangular asymmetrical wave V_M at its output. The ON time (Figure 5.4a) or OFF time (Figure 5.4b) of this rectangular wave V_M is given as

$$\delta_T = \frac{V_1}{V_R}T \qquad\qquad (5.7)$$

This rectangular wave V_M controls the switch S_2.

   (i) In Figure 5.4a, the switch S_2 is closed and connects another input voltage V_2 to the R_3C_2 low pass filter during this ON time δ_T. During the OFF time of V_M, the switch S_2 is opened and zero volts exist on the R_3C_2 low pass filter.

  (ii) In Figure 5.4b, the switch S_2 is opened and connects another input voltage V_2 to the R_3C_2 low pass filter during this OFF time δ_T. During the ON time of V_M, the switch S_2 is closed and zero volts exist on the R_3C_2 low pass filter.

In Figure 5.4a and b another asymmetrical pulse wave V_N is generated at the switch S_2 output. The maximum value of this pulse V_N is V_2 and the duty cycle is proportional to V_1. The R_3C_2 low pass filter gives the average value of V_N and is given as

$$V_O = \frac{1}{T}\int_0^{\delta_T} V_2\, dt = \frac{V_2}{T}\delta_T$$

$$V_O = \frac{V_1 V_2}{V_R} \qquad (5.8)$$

Design Exercises

1. The switches in Figure 5.4 can be replaced with the transistorized switches of Figure 1.13 (see Chapter 1) and shown in Figure 5.6. (i) Explain the working operation of the multipliers shown in Figure 5.6, (ii) draw waveforms at appropriate places and (iii) deduce expressions for their output voltages.

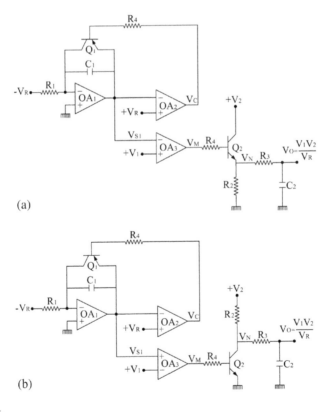

FIGURE 5.6
(a) Series switching saw tooth wave–based time division multiplier type II. (b) Shunt switching saw tooth wave–based time division multiplier type II.

2. The switches in Figure 5.4 are to be replaced with the FET switches shown in Figure 1.14 and the MOSFET switches shown in Figure 1.15 (see Chapter 1). In each, (i) draw the circuit diagrams, (ii) explain their working operations, (iii) draw waveforms at appropriate places and (iv) deduce expressions for their output voltages.

5.3 Triangular Wave–Referenced Time Division Multipliers: Type I

The circuit diagrams of triangular wave–based multipliers type I are shown in Figure 5.7 and their associated waveforms are shown in Figure 5.8.

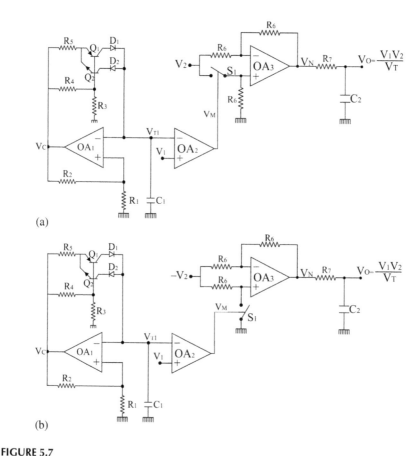

(a)

(b)

FIGURE 5.7
(a) Series switching triangular wave–based time division multiplier type 1. (b) Shunt switching triangular wave–based time division multiplier type I.

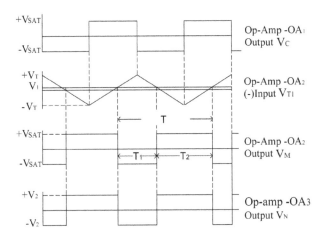

FIGURE 5.8
Associated waveforms of Figure 5.7.

Figure 5.7a shows a series switching TDM and 5.7b shows a shunt or parallel switching TDM. In Figure 5.7, as discussed in Chapter 3 (Section 3.2, Figure 3.3a), a triangular wave V_{T1} with $\pm V_T$ peak-to-peak values and time period T is generated by the op-amp OA_1 and transistors Q_1 and Q_2. The value of V_T is given as

$$V_T = \beta(V_{SAT}) \approx \beta(0.76)(V_{CC}) \tag{5.9}$$

where β is given as $\beta = R_1 / (R_1 + R_2)$.
   And time period T is given as

$$T = 4R_5C_1 \frac{R_1}{R_2}$$

One input voltage V_1 is compared with the generated triangular wave V_{T1} by the comparator OA_2. An asymmetrical rectangular waveform V_M is generated at the comparator OA_2 output. From the waveforms shown in Figure 5.8, it is observed that

$$T_1 = \frac{V_T - V_1}{2V_T} T$$

$$T_2 = \frac{V_T + V_1}{2V_T} T$$

$$T = T_1 + T_2 \tag{5.10}$$

The rectangular wave V_M controls the switch S_1.

1. First, let us consider the circuit shown in Figure 5.7a. During HIGH time (T_2) of the rectangular waveform V_M, the switch S_1 is closed, the op-amp OA_3 along with resistors R_6 will work as a non-inverting amplifier and $+V_2$ will appear at its output ($V_N = +V_2$). During LOW time (T_1) of the rectangular waveform V_M, the switch S_1 is opened, the op-amp OA_3 along with resistors R_6 will work as an inverting amplifier and $-V_2$ will appear at its output ($V_N = -V_2$).

2. Next, let us consider the circuit shown in Figure 5.7b. During HIGH time (T_2) of the rectangular waveform V_M, the switch S_1 is closed, the op-amp OA_3 along with resistors R_6 will work as an inverting amplifier and $+V_2$ will appear at its output ($V_N = +V_2$). During LOW time (T_1) of the rectangular waveform V_M, the switch S_1 is opened, the op-amp OA_3 along with resistors R_6 will work as a non-inverting amplifier and $-V_2$ will appear at its output ($V_N = -V_2$).

In both circuits shown in Figure 5.7, another rectangular wave V_N with peak-to-peak values of $\pm V_2$ is generated at the output of op-amp OA_3. The R_7C_2 low pass filter gives the average value of this pulse train V_N and is given as

$$V_O = \frac{1}{T}\left[\int_O^{T_2} V_2\,dt + \int_{T_2}^{T_1+T_2} (-V_2)\,dt\right] = \frac{V_2}{T}(T_2 - T_1) \tag{5.11}$$

Equation 5.10 in Equation 5.11 gives

$$V_O = \frac{V_1 V_2}{V_T} \tag{5.12}$$

Design Exercises

1. The switch S_1 in Figure 5.7 can be replaced with transistorized switches of Figure 1.13 (see Chapter 1) and shown in Figure 5.9. (i) Explain the working operation of the multipliers shown in Figure 5.9, (ii) draw waveforms at appropriate places and (iii) deduce expressions for their output voltages.

2. The switch S_1 in Figure 5.7 is to be replaced with the FET switches of Figure 1.14 and MOSFET switches of Figure 1.15 (see Chapter 1). In each, (i) draw the circuit diagrams, (ii) explain their working operations, (iii) draw waveforms at appropriate places and (iv) deduce expressions for their output voltages.

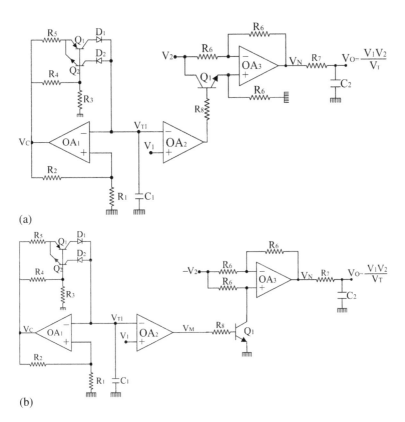

FIGURE 5.9
(a) Series switching triangular wave–based multiplier type I. (b) Shunt switching triangular wave–based multiplier type 1.

5.4 Triangular Wave–Based Time Division Multipliers: Type II

The circuit diagrams of triangular wave–referenced time division multipliers type II are shown in Figure 5.10 and their associated waveforms are shown in Figure 5.11. Figure 5.10a shows a series switching TDM and 5.10b shows a shunt or parallel switching TDM. As discussed in Chapter 3 (Section 3.2, Figure 3.3b), op-amps OA_1 and OA_2 constitute a triangular/square wave generator. The output of op-amp OA_1 is a triangular wave V_{T1} with $\pm V_T$ peak values and time period of T. The output of OA_2 is a square waveform with $\pm V_{SAT}$ peak-to-peak values:

$$\pm V_T = \pm V_{SAT}\frac{R_2}{R_3} \tag{5.13}$$

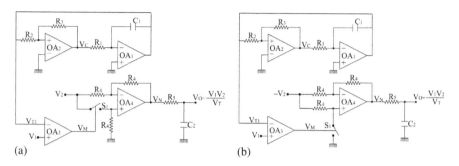

FIGURE 5.10
(a) Series switching triangular wave–based time division multiplier type II. (b) Parallel switching triangular wave–based time division multiplier type II.

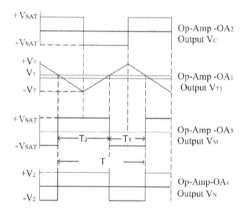

FIGURE 5.11
Associated waveforms of Figure 5.10.

$$T = 4R_1C_1 \frac{R_2}{R_3}$$

The comparator OA_3 compares this triangular wave V_{T1} with one input voltage V_1 and produces an asymmetrical rectangular wave V_M. From Figure 5.11, it is observed that

$$T_1 = \frac{V_T - V_1}{2V_T} T$$

$$T_2 = \frac{V_T + V_1}{2V_T} T$$

$$T = T_1 + T_2 \tag{5.14}$$

1. First, let us consider the circuit shown in Figure 5.10a. During HIGH time (T_2) of the rectangular waveform V_M, the switch S_1 is closed, the op-amp OA_4 along with resistors R_4 will work as a non-inverting amplifier and $+V_2$ will appear at its output ($V_N = +V_2$). During LOW time (T_1) of the rectangular waveform V_M, the switch S_1 is opened, the op-amp OA_4 along with resistors R_4 will work as an inverting amplifier and $-V_2$ will appear at its output ($V_N = -V_2$).

2. Next, let us consider the circuit shown in Figure 5.10b. During HIGH time (T_2) of the rectangular waveform V_M, the switch S_1 is closed, the op-amp OA_4 along with resistors R_4 will work as an inverting amplifier and $+V_2$ will appear at its output ($V_N = +V_2$). During LOW time (T_1) of the rectangular waveform V_M, the switch S_1 is opened, the op-amp OA_4 along with resistors R_4 will work as a non-inverting amplifier and $-V_2$ will appear at its output ($V_N = -V_2$).

In both the circuits shown in Figure 5.10, another asymmetrical rectangular waveform V_N is generated at the op-amp OA_4 output with $\pm V_2$ peak-to-peak values. The $R_5 C_2$ low pass filter gives the average value V_O and is given as

$$V_O = \frac{1}{T}\left[\int_O^{T_2} V_2\,dt + \int_{T_2}^{T_1+T_2} (-V_2)\,dt\right] = \frac{V_2}{T}\left[T_2 - T_1\right] \qquad (5.15)$$

Equation 5.14 in Equation 5.15 gives

$$V_O = \frac{V_1 V_2}{V_T} \qquad (5.16)$$

Design Exercises

1. The switch S_1 in Figure 5.10 can be replaced with the transistorized switches of Figure 1.13 (see Chapter 1) and shown in Figures 5.12. (i) Explain the working operation of the multipliers shown in Figures 5.12, (ii) draw waveforms at appropriate places and (iii) deduce expressions for their output voltages.

2. The switch S_1 in Figures 5.10 is to be replaced with the FET switches shown in Figure 1.14 and the MOSFET switches shown in Figure 1.15 (see Chapter 1). In each, (i) draw the circuit diagrams, (ii) explain their working operations, (iii) draw waveforms at appropriate places and (iv) deduce expressions for their output voltages.

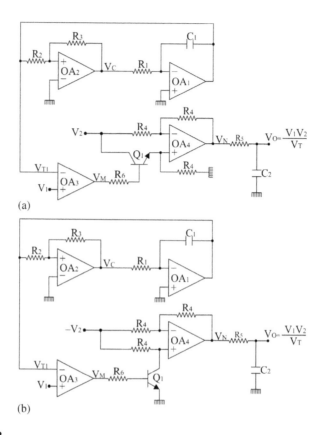

FIGURE 5.12
(a) Series switching triangular wave–based multiplier type II. (b) Shunt switching triangular wave–based multiplier type II.

5.5 Time Division Multipliers with No Reference: Type 1

The circuit diagrams of time division multipliers without using either a saw tooth wave or triangular wave as reference are shown in Figure 5.13 and their associated waveforms are shown in Figure 5.14. Figure 5.13a and b shows a series switching TDM and Figure 5.13c and d shows a shunt or parallel switching TDM. As discussed in Chapter 3 (Section 3.3, Figure 3.6), the op-amps OA_1 and OA_2 along with R_1, C_1, R_2 and R_3 constitute an asymmetrical rectangular wave generator, and a rectangular wave V_M is generated at the output of op-amp OA_2.

$$\pm V_T = \frac{R_2}{R_3} V_{SAT} \tag{5.17}$$

From waveforms shown in Figure 5.14, it is observed that

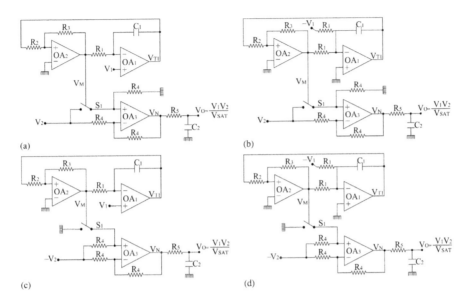

FIGURE 5.13
(a) Series switching time division multiplier with no clock type I. (b) Equivalent circuit of part (a). (c) Shunt switching time division with no reference type I. (d) Equivalent circuit of part (c).

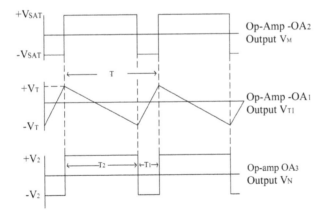

FIGURE 5.14
Associated waveforms of Figure 5.13.

$$T_1 = \frac{V_{SAT} - V_1}{2V_{SAT}} T$$

$$T_2 = \frac{V_{SAT} + V_1}{2V_{SAT}} T$$

$$T = T_1 + T_2 \tag{5.18}$$

1. Let us consider Figure 5.13a and b. During HIGH time (T_2) of the rectangular waveform V_M, the switch S_1 is closed, the op-amp OA_3 along with resistors R_4 will work as a non-inverting amplifier and $+V_2$ will appear at its output ($V_N = +V_2$). During LOW time (T_1) of the rectangular waveform V_M, the switch S_1 is opened, the op-amp OA_3 along with resistors R_4 will work as an inverting amplifier and $-V_2$ will appear at its output ($V_N = -V_2$).

2. Let us consider Figure 5.13c and d. During HIGH time (T_2) of the rectangular waveform V_M, the switch S_1 is closed, the op-amp OA_3 along with resistors R_4 will work as an inverting amplifier and $+V_2$ will appear at its output ($V_N = +V_2$). During LOW time (T_1) of the rectangular waveform V_M, the switch S_1 is opened, the op-amp OA_3 along with resistors R_4 will work as a non-inverting amplifier and $-V_2$ will appear at its output ($V_N = -V_2$).

In all circuits of Figure 5.13, another asymmetrical rectangular wave V_N is generated at the output of op-amp OA_3 with $\pm V_2$ as the maximum value. The R_5C_2 low pass filter gives the average value of this pulse train V_N and is given as

$$V_O = \frac{1}{T}\left[\int_O^{T_2} V_2\, dt + \int_{T_2}^{T_1+T_2} (-V_2)\, dt\right] = \frac{V_2}{T}[T_2 - T_1] \qquad (5.19)$$

Equation 5.18 in Equation 5.19 gives

$$V_O = \frac{V_1 V_2}{V_{SAT}} \qquad (5.20)$$

Design Exercises

1. The switch S_1 in Figure 5.13 can be replaced with transistorized switches of Figure 1.13 (see Chapter 1) and shown in Figure 5.15. (i) Explain the working operation of the multipliers shown in Figure 5.15, (ii) draw waveforms at appropriate places and (iii) deduce expressions for their output voltages.

2. The switch S_1 in Figure 5.13 is to be replaced with the FET switches shown in Figure 1.14 and the MOSFET switches shown in Figure 1.15. In each, (i) draw the circuit diagrams, (ii) explain their working operations, (iii) draw waveforms at appropriate places and (iv) deduce expressions for their output voltages.

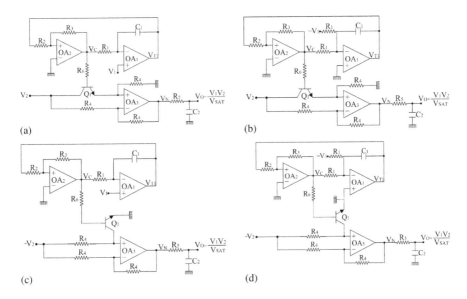

FIGURE 5.15
(a) Series switching TDM with no reference type I. (b) Equivalent circuit of part (a). (c) Shunt switching TDM with no reference type I. (d) Equivalent circuit of part (c).

5.6 Time Division Multipliers with No Reference: Type II

The circuit diagrams of time division multipliers type II without using a triangular or saw tooth wave as reference are shown in Figure 5.16 and their associated waveforms in Figure 5.17. Figure 5.16 shows a series switching TDM and Figure 5.16 shows a shunt or parallel switching TDM. As discussed in Chapter 3 (Section 3.4, Figure 3.9), the op-amps OA_1 and OA_2 along with R_1, C_1, R_2 and R_3 constitute an asymmetrical rectangular wave generator, and a rectangular wave V_C is generated at the output of op-amp OA_2.

$$\pm V_T = \pm \frac{R_2}{R_3} V_{SAT} \tag{5.21}$$

(i) First, let us consider the series switching multipliers shown in Figure 5.16a and b. During HIGH time (T_2) of V_C, the switch S_1 is closed and the op-amp OA_3 along with resistors R_4 will work as a non-inverting amplifier, and $+V_1$ will exist at its output ($V_M = +V_1$). During LOW time (T_1) of V_C, the switch S_1 is opened and the op-amp OA_3 along with resistors R_4 will work as an inverting amplifier, and $-V_1$ will exist at its output ($V_M = -V_1$).

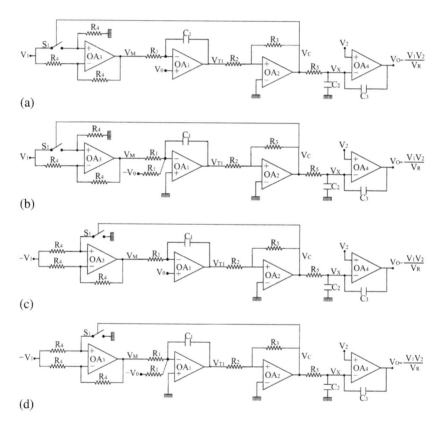

FIGURE 5.16

(a) Series switching time division multiplier with no reference type II. (b) Equivalent circuit of part (a). (c) Shunt switching time division multiplier with no reference type II. (d) Equivalent circuit of part (c).

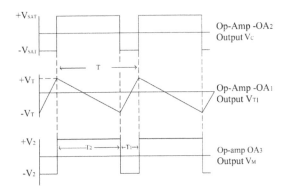

FIGURE 5.17
Associated waveforms of Figure 5.16.

(ii) Next, let us consider the shunt or parallel switching multipliers shown in Figure 5.16c and d. During HIGH time (T_2) of V_C, the switch S_1 is closed and the op-amp OA_3 along with resistors R_4 will work as an inverting amplifier, and $+V_1$ will exist at its output ($V_M = +V_1$). During LOW time (T_1) of V_C, the switch S_1 is opened and the op-amp OA_3 along with resistors R_4 will work as a non-inverting amplifier, and $-V_1$ will exist at its output ($V_M = -V_1$).

In all the circuits shown in Figure 5.16, a rectangular waveform V_M with $\pm V_1$ peak-to-peak values is generated at the output of op-amp OA_3.

From waveforms shown in Figure 5.17, it is observed that

$$T_1 = \frac{V_1 - V_O}{2V_1} T$$

$$T_2 = \frac{V_1 + V_O}{2V_1} T$$

$$T = T_1 + T_2 \tag{5.22}$$

The R_5C_2 low pass filter gives the average value of the rectangular wave V_C and is given as

$$V_X = \frac{1}{T}\left[\int_O^{T_2} V_{SAT}\,dt + \int_{T_2}^{T_1+T_2} (-V_{SAT})\,dt \right]$$

$$= \frac{V_{SAT}}{T}[T_2 - T_1] \tag{5.23}$$

Equation 5.22 in Equation 5.23 gives

$$V_X = \frac{V_O V_{SAT}}{V_1} \qquad (5.24)$$

The op-amp OA_4 is at a negative closed feedback configuration and a positive direct current (dc) voltage is ensured in the feedback loop. Hence its non-inverting terminal voltage must equal its inverting terminal voltage:

$$V_2 = V_X \qquad (5.25)$$

From Equations 5.24, 5.25 and 5.21

$$V_O = \frac{V_1 V_2}{V_{SAT}}$$

Let us assume $V_R = V_{SAT}$, then

$$V_O = \frac{V_1 V_2}{V_R} \qquad (5.26)$$

Design Exercises

1. The switches in Figure 5.16 can be replaced with the transistorized switches of Figure 1.13 (see Chapter 1) and shown in Figure 5.18. (i) Explain the working operation of the multipliers shown in Figure 5.18, (ii) draw waveforms at appropriate places and (iii) deduce expressions for their output voltages.

2. The switches in Figure 5.16a and b are to be replaced with the FET switches shown in Figure 1.14 and the MOSFET switches shown in Figure 1.15 (see Chapter 1). In each, (i) draw the circuit diagrams, (ii) explain their working operations, (iii) draw waveforms at appropriate places and (iv) deduce expressions for their output voltages.

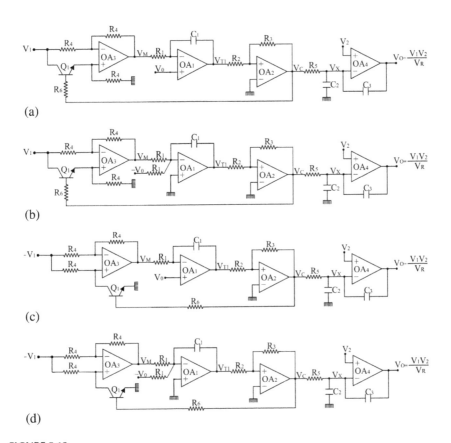

FIGURE 5.18
(a) Time division multiplier without reference type II. (b) Equivalent circuit of part (a). (c) Time division multiplier without reference type II. (d) Equivalent circuit of part (c).

Tutorial Exercises

5.1 Design series and shunt switching time division multipliers with a saw tooth wave generator.

5.2 You are given a triangular wave generator block. Design series- and shunt-type time division multipliers from this triangular wave generator. (i) Draw waveforms at appropriate places and (ii) deduce expressions for the output voltages.

5.3 In the multiplier circuit shown in Figure 5.7b, the comparator OA_2 terminals are interchanged and $-V_2$ is changed to $+V_2$. (i) Draw waveforms at appropriate places and (ii) deduce expressions for the output voltages.

5.4 In the multiplier circuits shown in Figures 5.1, 5.4, 5.7, 5.10, 5.13, and 5.16a and b, if the polarity of input voltage V_2 is reversed, (i) draw waveforms at appropriate places and (ii) deduce expressions their output voltages.

6

Peak Responding Multipliers (PRMs): Switching

As discussed in Chapter 4, peak responding multipliers are classified into (1) peak detecting multipliers and (2) peak sampling multipliers. A short pulse/saw tooth waveform whose time period (T) is proportional to one voltage is generated. Another input voltage is integrated during the time period. The peak value of the integrated voltage is proportional to the product of the input voltages. This is called a double single-slope peak responding multiplier. A square/triangular waveform whose time period is proportional to one voltage is generated. Another input voltage is integrated during the time period. The peak value of the integrated voltage is proportional to the product of the input voltages. This is called double dual-slope peak responding multiplier. A rectangular pulse waveform whose OFF time is proportional to one voltage is generated. Another voltage is integrated during this OFF time. The peak value of the integrated output is proportional to the product of the two input voltages. This is called a pulse width integrated peak responding multiplier.

At the output of a peak responding multiplier, if a peak detector is used, it is called a peak detecting multiplier, and if a sample and hold is used, it is called a sampling multiplier. A peak responding multiplier uses either analog switches or analog multiplexers for its operation. If analog switches are used, it is called a switching peak responding multiplier, and if analog multiplexers are used, it is called a multiplexing peak responding multiplier. Multiplexing peak responding multipliers are discussed in Chapter 4 and switching peak responding multipliers are discussed in this chapter.

6.1 Double Single-Slope Peak Responding Multipliers

The circuit diagrams of double single-slope peak responding multipliers are shown in Figure 6.1 and their associated waveforms are shown in Figure 6.2. Figure 6.1a shows a double single-slope peak detecting multiplier and Figure 6.1b shows a double single-slope peak sampling multiplier. When the op-amp OA_2 output is LOW, the switch S_1 opens and an integrator formed by resistor R_1, capacitor C_1 and op-amp OA_1 integrates the reference voltage $-V_R$. The integrated output will be

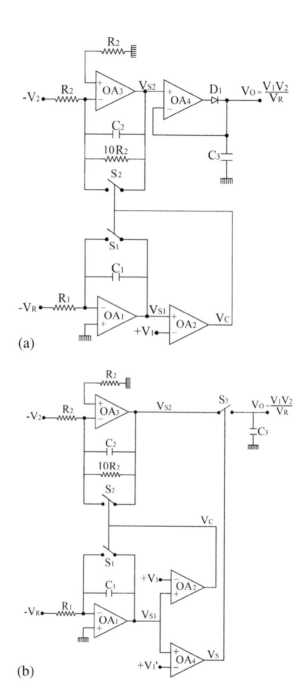

FIGURE 6.1
(a) Switching double single-slope peak detecting multiplier. (b) Switching double single-slope peak sampling multiplier.

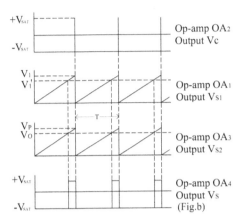

FIGURE 6.2
Associated waveforms of Figure 6.1.

$$V_{S1} = -\frac{1}{R_1C_1}\int -V_R dt = \frac{V_R}{R_1C_1}t \qquad (6.1)$$

A positive going ramp V_{S1} is generated at the output of op-amp OA_1. When the output of OA_1 reaches the voltage level of V_1, the comparator OA_2 output becomes HIGH. The switch S_1 is closed and hence the capacitor C_1 is shorted so that op-amp OA_1 output becomes zero. Then op-amp OA_2 output goes to LOW, the switch S_1 is opened and the integrator composed by R_1, C_1 and op-amp OA_1 integrates the reference voltage $-V_R$, and the cycle therefore repeats to provide (1) a saw tooth wave of peak value V_1 at the output of op-amp OA_1 and (2) a short pulse waveform V_C at the output of comparator OA_2. The short pulse V_C also controls switch S_2. During the short HIGH time of V_C, switch S_2 is closed and the capacitor C_2 is short-circuited so that op-amp OA_3 output is zero volts. During LOW time of V_C, switch S_2 is opened, the integrator formed by resistor R_2, capacitor C_2 and op-amp OA_3 integrates its input voltage $-V_2$ and its output is given as

$$V_{S2} = -\frac{1}{R_2C_2}\int -V_2 dt = \frac{V_2}{R_2C_2}t \qquad (6.2)$$

Another saw tooth waveform V_{S2} with peak value V_P is generated at the output of op-amp OA_3. From the waveforms shown in Figure 6.2, from Equations 6.1 and 6.2, and the fact that at $t = T$, $V_{S1} = V_1$, $V_{S2} = V_P$:

$$V_1 = \frac{V_R}{R_1C_1}T \qquad (6.3)$$

$$V_P = \frac{V_2}{R_2C_2}T \qquad (6.4)$$

From Equations 6.3 and 6.4

$$V_P = \frac{V_2}{R_2 C_2} \frac{V_1}{V_R} R_1 C_1$$

Let us assume $R_1 = R_2$ and $C_1 = C_2$, then

$$V_P = \frac{V_1 V_2}{V_R} \tag{6.5}$$

(i) In the circuit shown in Figure 6.1a, the peak detector realized by op-amp OA_4, diode D_1 and capacitor C_3 gives the peak value V_P at its output V_O: $V_O = V_P$.

(ii) In the circuit shown in Figure 6.1b, the peak value V_P is obtained by the sample and hold circuit realized by switch S_3 and capacitor C_3. The sampling pulse is generated by op-amp OA_4 by comparing a slightly less than voltage of V_1, called V_1', with the saw tooth wave V_{S1} by the comparator OA_4. The sample and hold operation is illustrated graphically in Figure 6.2. The sample and hold output is $V_O = V_P$.

From Equation 6.5, the output will be $V_O = V_P$

$$V_O = \frac{V_1 V_2}{V_R} \tag{6.6}$$

Design Exercises

1. The switches S_1 and S_2 in Figure 6.1 can be replaced with the transistorized switches of Figure 1.13 (see Chapter 1) and shown in Figure 6.3. (i) Explain the working operation of the multipliers shown in Figure 6.3, (ii) draw waveforms at appropriate places and (iii) deduce expressions for their output voltages.

2. The switches S_1 and S_2 in Figure 6.1 are to be replaced with the FET switches of Figure 1.14 and the MOSFET switches of Figure 1.15 (see Chapter 1). In each, (i) draw the circuit diagrams, (ii) explain their working operations, (iii) draw waveforms at appropriate places and (iv) deduce expressions for their output voltages.

6.2 Double Dual-Slope Peak Responding Multipliers: Type I

The circuit diagrams of double dual-slope peak responding multipliers type I are shown in Figure 6.4 and their associated waveforms are shown

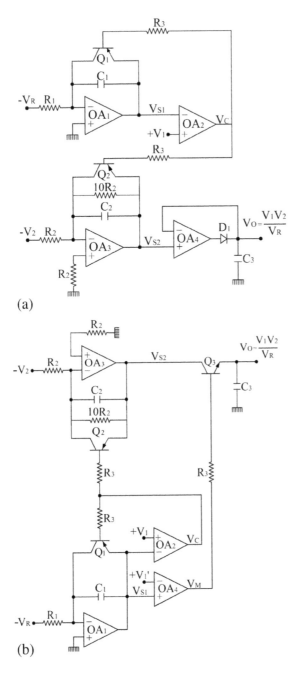

FIGURE 6.3
(a) Switching double single-slope peak detecting multiplier. (b) Switching double single-slope peak sampling multiplier.

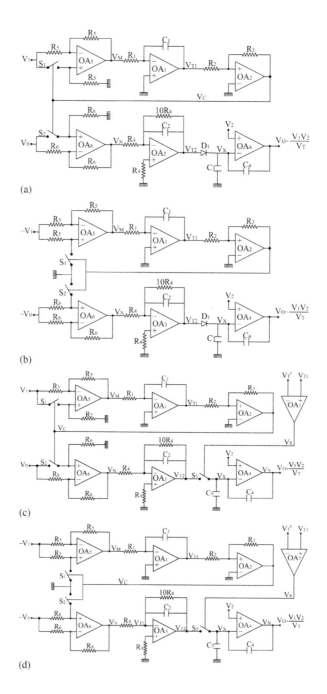

FIGURE 6.4
(a) Series switching double dual-slope peak detecting multiplier type I. (b) Parallel switching double dual-slope peak detecting multiplier type I. (c) Series switching double dual-slope peak sampling multiplier type I. (d) Shunt switching double dual-slope peak sampling multiplier type I.

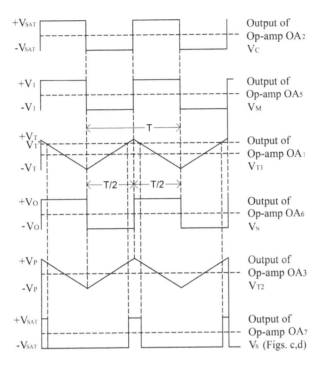

FIGURE 6.5
Associated waveforms of Figure 6.4.

in Figure 6.5. Figure 6.4a shows a series switching double dual-slope peak detecting multiplier, Figure 6.4b shows a shunt switching double dual-slope peak detecting multiplier, Figure 6.4c shows a series switching double dual-slope peak sampling multiplier and Figure 6.4d shows a shunt switching double-dual slope peak sampling multiplier. Let the comparator OA_2 output be LOW.

(i) In Figure 6.4a and c, the switch S_1 is opened, the op-amp OA_5 along with resistors R_5 will work as an inverting amplifier, and $-V_1$ will appear at its output ($V_M = -V_1$) and is given to the integrator formed by resistor R_1, capacitor C_1 and op-amp OA_1. The switch S_2 is opened, the op-amp OA_6 along with resistors R_6 will work as an inverting amplifier, and $-V_O$ will appear at its output ($V_N = -V_O$) and is given to the integrator formed by resistor R_4, capacitor C_2 and op-amp OA_3.

(ii) In Figure 6.4b and d, the switch S_1 is opened, the op-amp OA_5 along with resistors R_5 will work as a non-inverting amplifier, and $-V_1$ will appear at its output ($V_M = -V_1$) and is given to the integrator formed by resistor R_1, capacitor C_1 and op-amp OA_1. The switch S_2 is opened, the op-amp OA_6 along with resistors R_6 will work as a non-inverting amplifier, and $-V_O$ will appear at its output ($V_N = -V_O$) and is given to the integrator formed by resistor R_4, capacitor C_2 and op-amp OA_3.

The output of op-amp OA_1 will be

$$V_{T1} = -\frac{1}{R_1 C_1} \int -V_1 dt = \frac{V_1}{R_1 C_1} t \qquad (6.7)$$

The output of op-amp OA_3 will be

$$V_{T2} = -\frac{1}{R_4 C_2} \int -V_O dt = \frac{V_O}{R_4 C_2} t \qquad (6.8)$$

The output of op-amp OA_1 is a positive going ramp and when it reaches the value $+V_T$, the output of comparator OA_2 becomes HIGH.

(i) In Figure 6.4a and c, the switch S_1 is closed, the op-amp OA_5 along with resistors R_5 will work as a non-inverting amplifier, and $+V_1$ will appear at its output ($V_M = +V_1$) and is given to the integrator formed by resistor R_1, capacitor C_1 and op-amp OA_1. The switch S_2 is closed, the op-amp OA_6 along with resistors R_6 will work as a non-inverting amplifier, and $+V_O$ will appear at its output ($V_N = +V_O$) and is given to the integrator formed by resistor R_4, capacitor C_2 and op-amp OA_3.

(ii) In Figure 6.4b and d, the switch S_1 is closed, the op-amp OA_5 along with resistors R_5 will work as an inverting amplifier, and $+V_1$ will appear at its output ($V_M = +V_1$) and is given to the integrator formed by resistor R_1, capacitor C_1 and op-amp OA_1. The switch S_2 is closed, the op-amp OA_6 along with resistors R_6 will work as an inverting amplifier, and $+V_O$ will appear at its output ($V_N = +V_O$) and is given to the integrator formed by resistor R_4, capacitor C_2 and op-amp OA_3.

The output of op-amp OA_1 will now be

$$V_{T1} = -\frac{1}{R_1 C_1} \int V_1 dt = -\frac{V_1}{R_1 C_1} t \qquad (6.9)$$

The output of op-amp OA_3 will now be

$$V_{T2} = -\frac{1}{R_4 C_2} \int V_O dt = -\frac{V_O}{R_4 C_2} t \qquad (6.10)$$

The output of op-amp OA_1 changes slope from $+V_T$ toward $-V_T$, and when it reaches a value of $-V_T$, the comparator OA_2 becomes LOW and the cycle therefore repeats to give (1) a triangular waveform V_{T1} of $\pm V_T$ peak-to-peak values at the output of op-amp OA_1, (2) another triangular waveform V_{T2}

of $\pm V_P$ peak-to-peak values at the output of op-amp OA_3, (3) a first square waveform V_C with $\pm V_{SAT}$ peak-to-peak values at the output of op-amp OA_2, (4) a second square waveform V_M with $\pm V_1$ peak-to-peak values at the output of op-amp OA_5 and (5) a third square waveform V_N with $\pm V_O$ peak-to-peak values at the output of op-amp OA_6.

From the waveforms shown in Figure 6.5, Equations 6.7 and 6.8, and the fact that at $t = T/2$, $V_{T1} = 2V_T$, $V_{T2} = 2V_P$:

$$2V_T = \frac{V_1}{R_1 C_1} \frac{T}{2} \tag{6.11}$$

$$2V_P = \frac{V_O}{R_2 C_2} \frac{T}{2} \tag{6.12}$$

From Equations 6.11 and 6.12:

$$V_P = \frac{V_O V_T}{V_1} \frac{R_1 C_1}{R_2 C_2}$$

Let $R_1 = R_2$ and $C_1 = C_2$, then

$$V_P = \frac{V_O}{V_1} V_T \tag{6.13}$$

As discussed in Chapter 4 (Section 4.2, Figure 4.3)

$$\pm V_T = \pm V_{SAT} \frac{R_2}{R_3} \approx 0.76(\pm V_{CC}) \frac{R_2}{R_3} \tag{6.14}$$

In Figure 6.4a and b, the peak detector D_1 and capacitor C_3 give the peak value V_P of the triangular wave V_{T2}. Hence $V_X = V_P$.

(i) In Figure 6.4a and b, the peak detector D_1 and capacitor C_3 gives the peak value V_P of the triangular wave V_{T2}. Hence $V_X = V_P$.
(ii) In Figure 6.4c and d, the peak value V_P is obtained by the sample and hold circuit realized by switch S_3 and capacitor C_3. The sampling pulse is generated by op-amp OA_7 by comparing a slightly less than voltage of V_T, called V_T', with the triangular wave V_{T1} by the comparator OA_7. The sample and hold operation is illustrated graphically in Figure 6.5. The sample and hold output is $V_X = V_P$.

The op-amp OA_4 is kept in a negative closed loop configuration and a positive direct current (dc) voltage is ensured in the feedback. Hence its non-inverting terminal voltage will be equal to its inverting terminal voltage, i.e.,

$$V_2 = V_X = V_P \tag{6.15}$$

From Equations 6.13 and 6.15

$$V_O = \frac{V_1 V_2}{V_T} \tag{6.16}$$

Design Exercises

1. The switches S_1 and S_2 in Figure 6.4a and b can be replaced with the transistorized switches of Figure 1.13 (see Chapter 1) and shown in Figure 6.6. (i) Explain the working operation of the multipliers shown in Figure 6.6a, (ii) draw waveforms at appropriate places and (iii) deduce expressions for their output voltages.

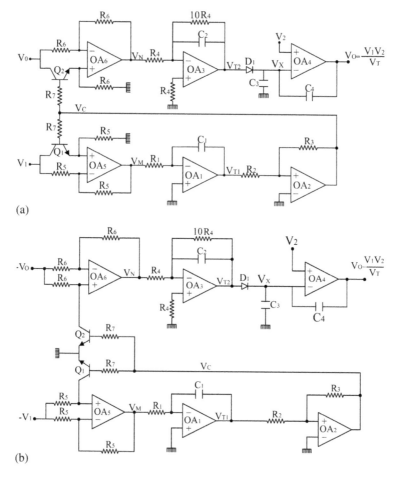

(a)

(b)

FIGURE 6.6
(a) Series switching double dual-slope peak detecting multiplier type I. (b) Shunt switching double dual-slope peak detecting multiplier type I.

2. The switches S_1 and S_2 in Figure 6.4a and b are to be replaced with the FET switches of Figure 1.14 and the MOSFET switches of Figure 1.15. In each, (i) draw the circuit diagrams, (ii) explain their working operations, (iii) draw waveforms at appropriate places and (iv) deduce expressions for their output voltages.

6.3 Double Dual-Slope Peak Responding Multipliers: Type II

The circuit diagrams of double dual-slope peak responding multipliers type II are shown in Figure 6.7 and their associated waveforms are shown in Figure 6.8. Figure 6.7a shows a series switching peak detecting multiplier, Figure 6.7b shows a shunt or parallel switching peak detecting multiplier, Figure 6.7c shows a series switching peak sampling multiplier and Figure 6.7d shows a shunt or parallel switching peak sampling multiplier. Let the comparator OA_2 output be LOW ($-V_{SAT}$).

(i) In Figure 6.7a and c, the switch S_1 is opened, the op-amp OA_5 along with resistors R_5 will work as a inverting amplifier and $-V_1$ will appear at its output ($V_M = -V_1$). The switch S_2 is opened, the op-amp OA_6 along with resistors R_6 will work as an inverting amplifier and $-V_2$ will appear at its output ($V_N = -V_2$).

(ii) In Figure 6.7b and d, the switch S_1 is opened, the op-amp OA_5 along with resistors R_5 will work as a non-inverting amplifier and $-V_1$ will appear at its output ($V_M = -V_1$). The switch S_2 is opened, the op-amp OA_6 along with resistors R_6 will work as a non-inverting amplifier and $-V_2$ will appear at its output ($V_N = -V_2$).

(iii) In all the circuits of Figure 6.7, $-V_{SAT}$ is given to the integrator formed by resistor R_1, capacitor C_1 and op-amp OA_1.

The op-amp OA_1 output is given as

$$V_{T1} = -\frac{1}{R_1 C_1} \int -V_{SAT}\, dt = \frac{V_{SAT}}{R_1 C_1} t \qquad (6.17)$$

The op-amp OA_3 output is given as

$$V_{T2} = -\frac{1}{R_4 C_2} \int -V_2\, dt = \frac{V_2}{R_4 C_2} t \qquad (6.18)$$

The output of op-amp OA_1 is rising toward positive saturation and when it reaches a value of $+V_T$, the comparator OA_2 output becomes HIGH ($+V_{SAT}$).

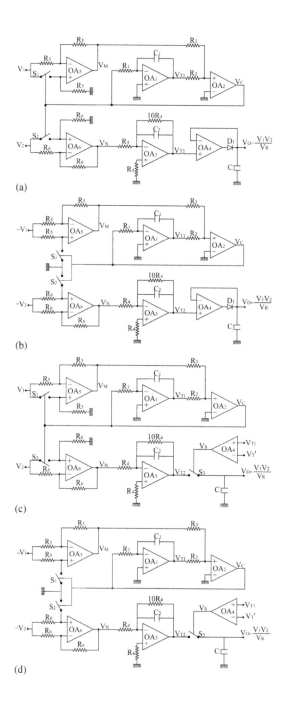

FIGURE 6.7
(a) Series switching double dual-slope multiplier type II. (b) Parallel switching double dual-slope multiplier type II. (c) Series switching double dual-slope peak sampling multiplier type II. (d) Shunt switching double dual-slope peak sampling multiplier type II.

(i) In Figure 6.7a and c, the switch S_1 is closed, the op-amp OA_5 along with resistors R_5 will work as a non-inverting amplifier and $+V_1$ will appear at its output ($V_M = +V_1$). The switch S_2 is closed, the op-amp OA_6 along with resistors R_6 will work as a non-inverting amplifier and $+V_2$ will appear at its output ($V_N = +V_2$).

(ii) In Figure 6.7b and d, the switch S_1 is closed, the op-amp OA_5 along with resistors R_5 will work as an inverting amplifier and $+V_1$ will appear at its output ($V_M = +V_1$). The switch S_2 is closed, the op-amp OA_6 along with resistors R_6 will work as an inverting amplifier and $+V_2$ will appear at its output ($V_N = +V_2$).

(iii) In all the circuits of Figure 6.7, $+V_{SAT}$ is given to the integrator formed by R_1, C_1 and op-amp OA_1.

The output of OA_1 will now be

$$V_{T1} = -\frac{1}{R_1C_1}\int V_{SAT}\,dt = -\frac{V_{SAT}}{R_1C_1}t \tag{6.19}$$

And the output of OA_3 will be

$$V_{T2} = -\frac{1}{R_4C_2}\int V_2\,dt = -\frac{V_2}{R_4C_2}t \tag{6.20}$$

The output of op-amp OA_1 changes its slope and moves toward negative saturation. When the output of op-amp integrator OA_1 comes down to a value $(-V_T)$, the comparator OA_2 output will become LOW and therefore the cycle repeats itself to give (1) a triangular wave V_{T1} at the output of op-amp OA_1 with $\pm V_T$ peak-to-peak values, (2) another triangular wave V_{T2} at the output of op-amp OA_3 with $\pm V_P$ peak-to-peak values, (3) a first square waveform V_C with $\pm V_{SAT}$ peak-to-peak values at the output of op-amp OA_2, (4) a second square waveform V_M with $\pm V_1$ as peak-to-peak values at the output of op-amp OA_5 and (5) a third square waveform V_N with $\pm V_2$ as peak-to-peak values at the output of op-amp OA_6. From the waveforms shown in Figure 6.8, Equations 6.17 and 6.18, and the fact that at $t = T/2$, $V_{T1} = 2V_T$, $V_{T2} = 2V_P$:

$$2V_T = \frac{V_{SAT}}{R_1C_1}\frac{T}{2} \tag{6.21}$$

$$2V_P = \frac{V_2}{R_4C_2}\frac{T}{2} \tag{6.22}$$

From Equations 6.21 and 6.22 by assuming $R_1 = R_4$ and $C_1 = C_2$:

$$V_P = \frac{V_2}{V_{SAT}}V_T$$

As discussed in Chapter 4 (Section 4.3, Figure 4.6)

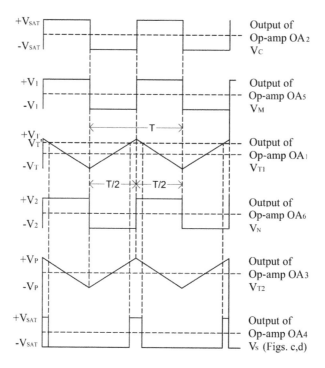

FIGURE 6.8
Associated waveforms of Figure 6.7.

$$\pm V_T = \pm \frac{R_2}{R_3} V_1$$

$$V_P = \frac{V_1 V_2}{V_{SAT}} \frac{R_2}{R_3}$$

Let

$$V_R = \frac{V_{SAT}}{R_2} R_3$$

$$V_P = \frac{V_1 V_2}{V_R} \qquad (6.23)$$

(i) In the circuits shown in Figure 6.7a and b, the peak detector realized by op-amp OA_4, diode D_1 and capacitor C_3 gives the peak value V_p at its output: $V_O = V_P$.

(ii) In the circuits shown in Figure 6.7c and d, the peak value V_P is obtained by the sample and hold circuit realized by switch S_3 and capacitor C_3. The sampling pulse V_S is generated by op-amp OA_4 by comparing a slightly less than voltage of V_T called V_T' with the triangular wave V_{T1}. The sample and hold operation is illustrated graphically in Figure 6.8. The sampled output is given as $V_O = V_P$.

From Equation 6.23 the output voltage will be $V_O = V_P$:

$$V_O = \frac{V_1 V_2}{V_R} \tag{6.24}$$

Design Exercises

1. The switches S_1 and S_2 in Figure 6.7a and b can be replaced with the transistorized switches of Figure 1.13 (see Chapter 1) and shown in Figure 6.9. (i) Explain the working operation of the multipliers shown in Figure 6.6, (ii) draw waveforms at appropriate places and (iii) deduce expressions for their output voltages.

2. The switches S_1 and S_2 in Figure 6.7a and b are to be replaced with the FET switches of Figure 1.14 and the MOSFET switches of Figure 1.15 (see Chapter 1). In each, (i) draw the circuit diagrams, (ii) explain their working operations, (iii) draw waveforms at appropriate places and (iv) deduce expressions for their output voltages.

6.4 Pulse Width Integrated Peak Responding Multipliers

The circuit diagrams of pulse width integrated peak responding multipliers are shown in Figure 6.10 and their associated waveforms are shown in Figure 6.11. Figure 6.10a shows a peak detecting multiplier and Figure 6.10b shows a peak sampling multiplier. When the comparator OA_2 output is LOW, the switch S_1 is opened and the integrator formed by resistor R_1, capacitor C_1 and op-amp OA_1 integrates $-V_R$. The integrated output is given as

$$V_{S1} = -\frac{1}{R_1 C_1} \int -V_R \, dt = \frac{V_R}{R_1 C_1} t \tag{6.25}$$

When the output of op-amp OA_1 is rising toward positive saturation and it reaches the value $+V_R$, the comparator OA_2 output will become HIGH, the switch S_1 is closed and the capacitor C_1 is short-circuited and op-amp OA_1 output becomes zero. Now the comparator OA_2 output changes to LOW and the cycle therefore repeats to give (1) a saw tooth waveform V_{S1} of peak value V_R and time period T at the output of op-amp OA_1 and (2) a short pulse

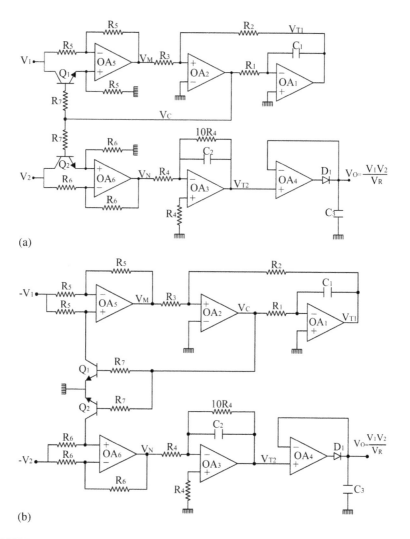

FIGURE 6.9
(a) Series switching double dual-slope multiplier type II. (b) Parallel switching double dual-slope multiplier type II.

waveform V_C at the output of op-amp OA_2. From the waveforms shown in Figure 6.11, and the fact that at $t = T$, $V_{S1} = +V_R$:

$$V_R = \frac{V_R}{R_1 C_1} T, \quad T = R_1 C_1 \tag{6.26}$$

The saw tooth waveform V_{S1} is compared with one input voltage V_1 by the comparator OA_3. An asymmetrical rectangular wave V_M is generated at the output of comparator OA_3. The OFF time of this wave is given as

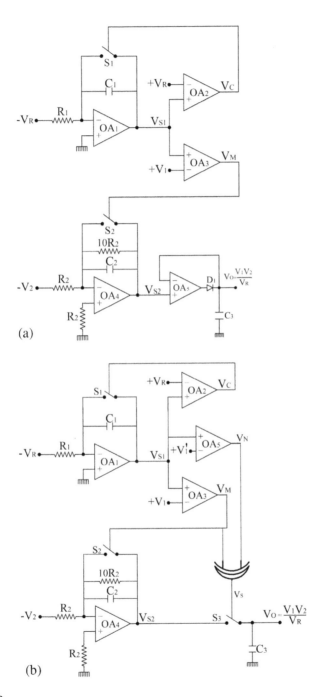

FIGURE 6.10
(a) Pulse width integrated peak detecting multiplier. (b) Pulse width integrated peak sampling multiplier.

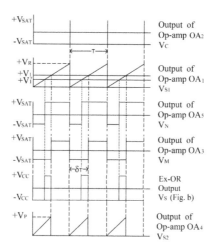

FIGURE 6.11
Associated waveforms of Figure 6.10.

$$\delta_T = \frac{V_1}{V_R} T \tag{6.27}$$

The output of comparator OA_3 is given as a control input of switch S_2. During the ON time of V_M, the switch S_2 is closed and the capacitor C_2 is shorted so that zero volts appear at op-amp OA_4 output. During the OFF time of V_M, the switch S_2 is opened and another integrator is formed by resistor R_2, capacitor C_2 and op-amp OA_4. This integrator integrates the input voltage $-V_2$ and its output is given as

$$V_{S2} = -\frac{1}{R_2 C_2} \int -V_2 \, dt = \frac{V_2}{R_2 C_2} t \tag{6.28}$$

A semi saw tooth wave V_{S2} with peak values of V_P is generated at the output of op-amp OA_4. From the waveforms shown in Figure 6.11, Equation 6.28, and the fact that at $t = \delta_T$, $V_{S2} = V_P$:

$$V_P = \frac{V_2}{R_2 C_2} \delta_T$$

$$V_P = \frac{V_1 V_2}{V_R} \frac{R_1 C_1}{R_2 C_2}$$

Let $R_1 = R_2$ and $C_1 = C_2$, then

$$V_P = \frac{V_1 V_2}{V_R} \tag{6.29}$$

(i) In the circuit shown in Figure 6.10a, the peak detector realized by op-amp OA_5, diode D_1 and capacitor C_3 gives this peak value V_P at its output: $V_0 = V_P$.

(ii) In the circuit shown in Figure 6.10b, the peak value V_P is obtained by the sample and hold circuit realized by switch S_3 and capacitor C_3. The sampling pulse V_S is generated by the Ex-OR gate from the signals V_M and V_N. V_N is obtained by comparing slightly less than voltage of V_1, i.e., V_1' with the sawtooth waveform V_{S1}. The sampled output is given as $V_0 = V_P$.

From Equation 6.29, the output voltage will be $V_0 = V_P$:

$$V_O = \frac{V_1 V_2}{V_R} \qquad (6.30)$$

Design Exercises

1. The switches S_1 and S_2 in Figure 6.10 can be replaced with the transistorized switches of Figure 1.13 (see Chapter 1) and shown in Figure 6.12. (i) Explain the working operation of the multipliers shown in Figure 6.12, (ii) draw waveforms at appropriate places and (iii) deduce expressions for their output voltages.

2. The switches S_1 and S_2 in Figure 6.10 are to be replaced with the FET switches of Figure 1.14 and the MOSFET switches of Figure 1.15 (see Chapter 1). In each, (i) draw the circuit diagrams, (ii) explain their working operations, (iii) draw waveforms at appropriate places and (iv) deduce expressions for their output voltages.

Tutorial Exercises

6.1 Design a pulse width integrated peak detecting multiplier using a saw tooth wave generator.

6.2 In the multiplier circuit shown in Figure 6.4a, find the value of V_X if the output voltage V_O is measured to be 10 V, $R_2 = 10$ KΩ, $R_3 = 15$ KΩ and $V_1 = 10$ V.

6.3 In the multiplier circuit shown in Figure 6.7b, if the polarity of diode D_1 is reversed, find out the output voltage.

6.4 In the multiplier circuit shown in Figure 6.7c, find the peak value of triangular wave V_{T2}, if $R_2 = 10$ KΩ, $R_3 = 15$ KΩ and $V_2 = 10$ V.

6.5 In the multiplier circuits shown in Figures 6.1, 6.4a and b, 6.7a and b, 6.10, if the polarity of the input voltage V_2 and direction of diode

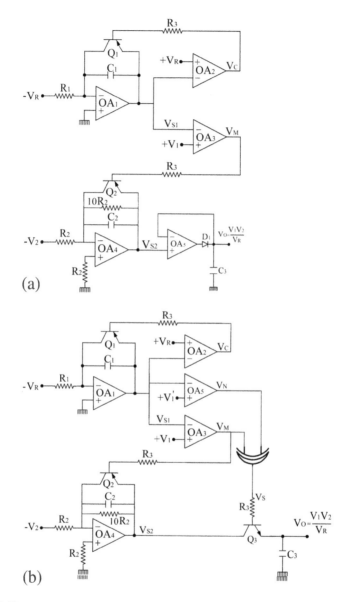

FIGURE 6.12

(a) Pulse width integrated peak detecting multiplier. (b) Pulse width integrated peak sampling multiplier.

D_1 are reversed, (i) draw waveforms at appropriate places and (ii) deduce expressions of their output voltages.

6.6 In the multiplier circuits shown in Figures 6.13 and 6.14, (i) deduce expressions for the output voltages and (ii) draw waveforms at appropriate places.

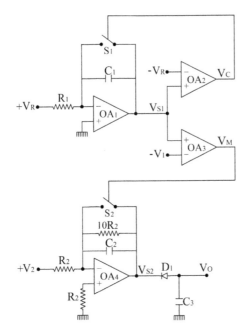

FIGURE 6.13
Multiplier circuit for Tutorial Exercise 6.6.

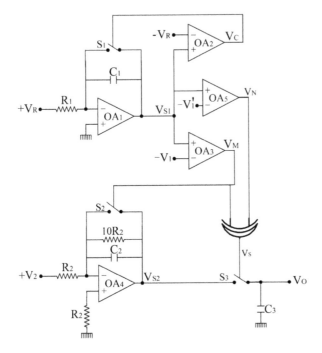

FIGURE 6.14
Multiplier circuit for Tutorial Exercise 6.6.

7

Pulse Position Responding Multipliers (PPRMs)

Pulse position responding multipliers (PPRMs) are classified into pulse position detecting multipliers (PPDMs) and pulse position sampling multipliers (PPSMs). A saw tooth wave of period T whose peak value is proportional to one input voltage is detected by a peak detector during a time δ_T, which is proportional to another input voltage. The peak detector output is proportional to the product of the two input voltages. δ_T is obtained from T and $\delta_T < T$. This is called a pulse position peak detecting multiplier.

A saw tooth wave of period T whose peak value is proportional to one input voltage is sampled by a sampling pulse whose position over the period T is proportional to another input voltage. The sampled output is proportional to the product of two input voltages. This is called a pulse position sampled multiplier. These two types of pulse position responding multipliers are discussed in this chapter.

7.1 Pulse Position Peak Responding Multipliers: Multiplexing Type

The circuit diagrams of multiplexing-type pulse position responding multipliers are shown in Figure 7.1 and their associated waveforms are shown in Figure 7.2. Figure 7.1a–c shows pulse position detecting multipliers and Figure 7.1d–f shows pulse position sampling multipliers.

In Figure 7.1a and d, as discussed in Chapter 3, Figure 3.1a, the op-amp OA_1, transistor Q_1 and multiplexer M_1 constitute a saw tooth wave generator. It produces (1) a saw tooth wave V_{S1} with peak value of V_R and (2) a short pulse wave V_C, at its outputs. The time period T of both the saw tooth V_{S1} and short pulse waveforms is given as

$$T = 1.4R_1C_1$$

$$V_R = 2V_{BE} \tag{7.1}$$

In Figure 7.1b and e, as discussed in Chapter 3, Figure 3.1b, the op-amps OA_1 and OA_2, and multiplexer M_1 constitute a saw tooth wave generator. It

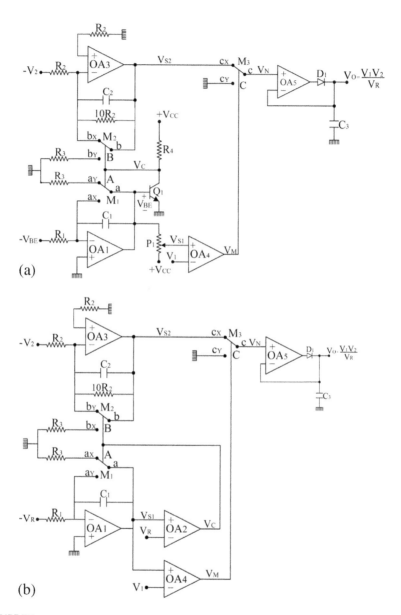

(a)

(b)

FIGURE 7.1

produces (1) a saw tooth wave V_{S1} with peak value of V_R and (2) a short pulse wave V_C at its outputs. The time period T of both the saw tooth V_{S1} and short pulse waveform V_C is given as

$$T = R_1 C_1 \tag{7.2}$$

The peak value V_R is the reference voltage V_R.

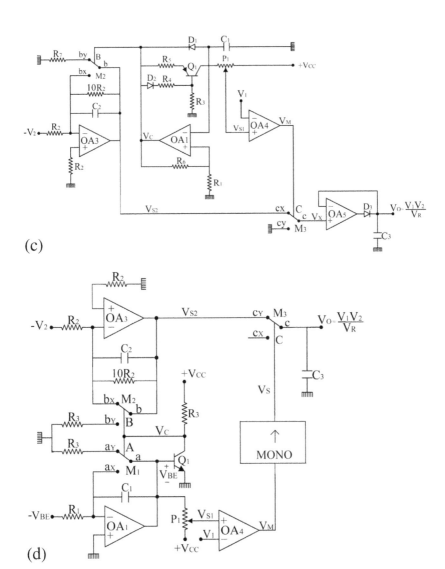

(c)

(d)

FIGURE 7.1 (CONTINUED)

In Figure 7.1c and f, as discussed in Chapter 3, Figure 3.1c, the op-amp OA_1; transistor Q_1; resistors R_1, R_6, R_3, R_4 and R_5; diodes D_1 and D_2; and capacitor C_1 constitute a saw tooth wave generator. It produces (1) a saw tooth wave V_{S1} with peak value of V_R and (2) a short pulse wave V_C at its outputs. The peak value V_R and the time period T of both the saw tooth V_{S1} and short pulse waveform V_C is given as

$$T = 2R_5C_1 \ln\left(1 + 2\frac{R_1}{R_6}\right)$$

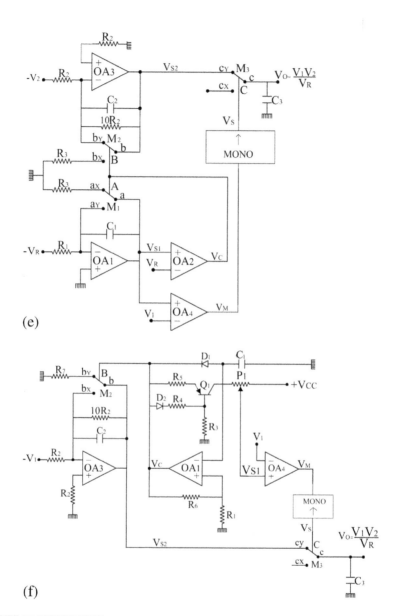

FIGURE 7.1 (CONTINUED)
(a) Pulse position peak detecting multiplier type I. (b) Pulse position detecting multiplier type II. (c) Pulse position detecting multiplier type II. (d) Pulse position peak sampling multiplier type I. (e) Pulse position sampled multiplier type II. (f) Pulse position sampling multiplier type III.

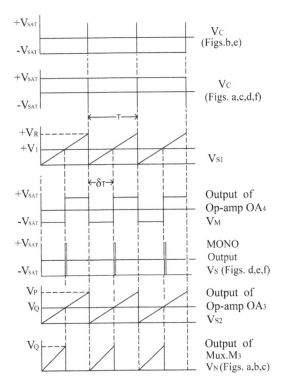

FIGURE 7.2
Associated waveforms of Figure 7.1a–d.

$$V_R = \beta(V_{SAT}) + \frac{\beta(V_{SAT})}{1.5} \qquad (7.3)$$

where β is given as $\beta = \dfrac{R_1}{R_1 + R_6}$.

In all circuits of Figure 7.1, the short pulse V_C is also given to multiplexer M_2, which constitutes a controlled integrator along with op-amp OA_3, resistor R_2 and capacitor C_2.

1. In Figure 7.1b and e, during HIGH value of V_C, the multiplexer M_2 connects 'by' to 'b', and capacitor C_2 is short-circuited so that integrator OA_3 output becomes zero. During LOW value of V_C, the multiplexer M_2 connects 'bx' to 'b', and another integrator is formed by op-amp OA_3, resistor R_2 and capacitor C_2.

2. In Figure 7.1a, c, d and f, during HIGH value of V_C, the multiplexer M_2 connects 'by' to 'b', and another integrator is formed by op-amp OA_3, resistor R_2 and capacitor C_2. During LOW value of V_C, the multiplexer M_2 connects 'bx' to 'b', and capacitor C_2 is short-circuited so that integrator OA_3 output becomes zero.

In all circuits of Figure 7.1, the integrator OA$_3$ output is given as

$$V_{S2} = -\frac{1}{R_2 C_2} \int -V_2 dt = \frac{V_2}{R_2 C_2} t \qquad (7.4)$$

Another saw tooth wave V_{S2} with peak value of V_P is generated at the output of integrator OA$_3$. From the waveforms shown in Figure 7.2, Equation 7.4, and the fact that at $t = T$, $V_{S2} = V_P$:

$$V_P = \frac{V_2}{R_2 C_2} T \qquad (7.5)$$

In all circuits of Figure 7.1, the comparator OA$_4$ compares the saw tooth wave V_{S1} with an input voltage V_1 and produces a rectangular wave V_M. The OFF time δ_T of this rectangular V_M is given as

$$\delta_T = \frac{V_1}{V_R} T \qquad (7.6)$$

1. In Figure 7.1a–c, the rectangular pulse V_M controls multiplexer M$_3$. During LOW value of V_M, the multiplexer M$_3$ connects 'cx' to 'c' and the saw tooth wave V_{S2} is connected to the multiplexer M$_3$ output. During HIGH value of V_M, the multiplexer M$_3$ connects 'cy' to 'c' and zero volts is connected to the multiplexer M$_3$ output. A semi saw tooth wave V_N with peak value V_Q is generated at the output of multiplexer M$_3$. The peak detector realized by op-amp OA$_5$, diode D$_1$ and capacitor C$_3$ gives this peak value V_Q at its output, i.e., $V_O = V_Q$.

2. In Figure 7.1d–f, the rectangular pulse V_M is given to the monostable multivibrator (MONO), which gives a narrow short spike during every rising edge of V_M. The MONO output V_S is acting as a sampling pulse to the sample and hold circuit realized by multiplexer M$_3$ and capacitor C$_3$. As shown in the waveforms in Figure 7.2, the sample and hold output is V_Q, i.e., $V_O = V_Q$.

The peak value V_Q is given as

$$V_Q = \frac{V_P}{T} \delta_T \qquad (7.7)$$

From Equations 7.5, 7.6 and 7.7:

$$V_O = \frac{V_1 V_2}{V_R} \frac{T}{R_2 C_2}$$

Let $T = R_2 C_2$, then

$$V_O = \frac{V_1 V_2}{V_R} \qquad (7.8)$$

Design Exercises

1. In the multiplier circuits shown in Figure 7.1, if the saw tooth wave generator is replaced with the saw tooth wave generator of Figure 3.1d, (i) draw the circuit diagrams, (ii) explain their working operations, (iii) draw waveforms at appropriate places and (iv) deduce expressions for their output voltages.

2. In the multiplier circuits shown in Figure 7.1, if the polarity of input voltage V_2 and direction of diode D_1 are reversed, (i) draw waveforms at appropriate places and (ii) deduce expressions for their output voltages.

7.2 Pulse Position Peak Responding Multipliers: Switching Type

The circuit diagrams of switching-type pulse position peak responding multipliers are shown in Figure 7.3 and their associated waveforms are shown in Figure 7.4. Figure 7.3a shows a pulse position peak detecting multiplier and Figure 7.3b shows a pulse position sampling multiplier. When the op-amp OA_2 output is LOW, the switch S_1 opens, and an integrator formed by resistor R_1, capacitor C_1 and op-amp OA_1 integrates the reference voltage $-V_R$. The integrator output is given as

$$V_{S1} = -\frac{1}{R_1 C_1} \int -V_R dt = \frac{V_R}{R_1 C_1} t \qquad (7.9)$$

A positive going ramp V_{S1} is generated at the output of op-amp OA_1. When the output of OA_1 reaches the voltage level of $+V_R$, the comparator OA_2 output becomes HIGH. The switch S_1 is closed and hence the capacitor C_1 is shorted so that op-amp OA_1 output becomes ZERO. Then op-amp OA_2 output goes to LOW, the switch S_1 opens and the integrator composed by R_1, C_1 and op-amp OA_1 integrates the reference voltage $-V_R$ and the cycle therefore repeats to provide (1) a saw tooth wave of peak value V_R at the output of op-amp OA_1 and (2) a short pulse waveform V_C at the output of comparator OA_2. The short pulse V_C also controls switch S_2. During the short HIGH time of V_C, switch S_2 is closed, and the capacitor C_2 is short-circuited so that op-amp OA_3 output is zero volts. During LOW time of V_C, switch S_2 opens, and the integrator formed by resistor R_2, capacitor C_2, and op-amp OA_3 integrates its input voltage $-V_2$ and its output is given as

$$V_{S2} = -\frac{1}{R_2 C_2} \int -V_2 \, dt = \frac{V_2}{R_2 C_2} t \qquad (7.10)$$

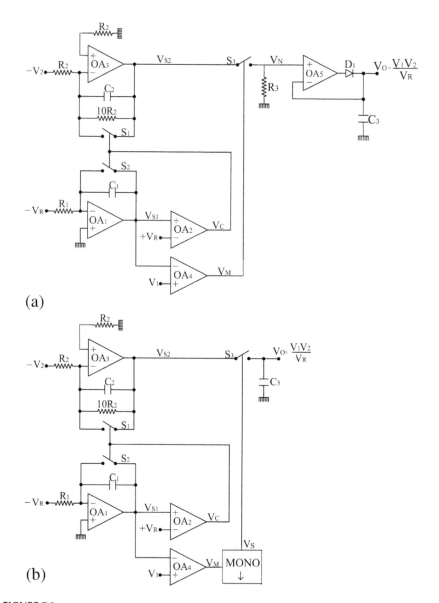

(a)

(b)

FIGURE 7.3
(a) Pulse position detecting multiplier. (b) Pulse position sampled multiplier.

Another saw tooth waveform V_{S2} with peak value V_P is generated at the output of op-amp OA_3. From the waveforms shown in Figure 7.4, from Equations 7.9 and 7.10, and the fact that at $t = T$, $V_{S1} = V_R$, $V_{S2} = V_P$:

$$V_R = \frac{V_R}{R_1 C_1} T, \quad T = R_1 C_1$$

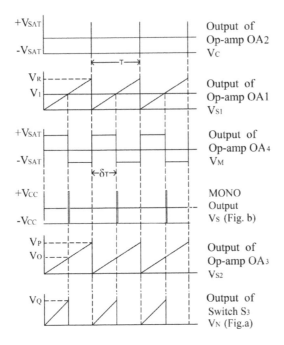

FIGURE 7.4
Associated waveforms of Figure 7.3.

$$V_P = \frac{V_2}{R_2C_2}T, \quad V_P = \frac{V_2}{R_2C_2}R_1C_1$$

Let us assume $R_1 = R_2$ and $C_1 = C_2$, then

$$V_P = V_2 \qquad (7.11)$$

The saw tooth wave V_{S1} is compared with the first input voltage V_1 by the comparator OA_4 and a rectangular wave V_M is generated at the output of comparator OA_4.

The ON time of this rectangular waveform V_M is given as

$$\delta_T = \frac{V_1}{V_R}T \qquad (7.12)$$

1. In Figure 7.3a, the rectangular pulse V_M controls the switch S_3. During HIGH time of V_M, the switch S_3 is closed and the saw tooth wave V_{S2} is connected to the peak detector realized by op-amp OA_5, diode D_1 and capacitor C_3. During LOW time of V_M, the switch S_3 is opened and zero volts exist on the peak detector. A semi saw tooth wave V_N with peak value V_Q is generated at the output of switch

S_3. The peak detector gives this peak value V_Q at its output. Hence $V_O = V_Q$.

2. In Figure 7.3b, the rectangular waveform V_M is given to the monostable multivibrator (MONO), which gives a narrow short pulse during every falling edge of V_M. That is, a short pulse V_S is generated during the time δ_T. This short pulse V_S is used as a sampling pulse to the sample and hold circuit realized by switch S_3 and capacitor C_3. The sample and hold output is $V_O = V_Q$.

The peak value V_Q is given as

$$V_Q = \frac{V_P}{T}\delta_T \qquad (7.13)$$

Equations 7.11 and 7.12 in Equation 7.13 gives

$$V_O = \frac{V_1 V_2}{V_R} \qquad (7.14)$$

Design Exercises

1. The switch S_3 in Figure 7.3 is to be replaced with the transistorized switches of Figure 1.13 (see Chapter 1). (i) Draw circuit diagrams, (ii) explain their working operation, (iii) draw waveforms at various points and (iv) deduce expressions for their output voltages.

2. The switch S_3 in Figure 7.3 are to be replaced with the FET switches of Figure 1.14 and the MOSFET switches of Figure 1.15 (see Chapter 1). In each, (i) draw the circuit diagrams, (ii) explain their working operations, (iii) draw waveforms at appropriate places and (iv) deduce expressions for their output voltages.

7.3 Pulse Position Peak Detecting Multipliers with Transistors

The circuit diagrams of pulse position peak detecting multipliers with transistors are shown in Figure 7.5 and their associated waveforms are shown in Figure 7.6.

In Figure 7.5a, when the op-amp OA_2 output is HIGH, the transistor Q_1 is OFF and an integrator formed by resistor R_1, capacitor C_1 and op-amp OA_1 integrates the reference voltage $-V_R$. The integrator output is given as

$$V_{S1} = -\frac{1}{R_1 C_1}\int -V_R\, dt = \frac{V_R}{R_1 C_1}t \qquad (7.15)$$

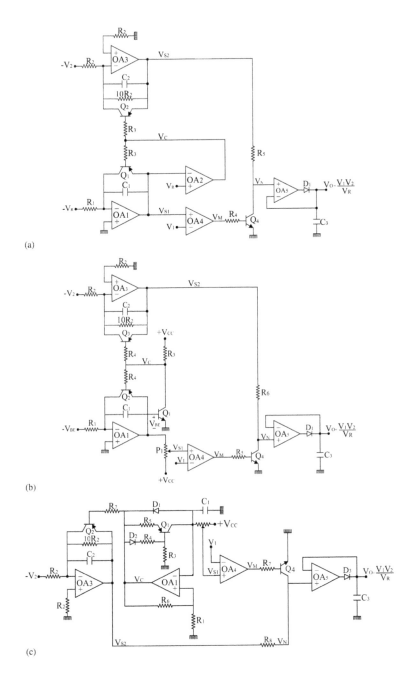

FIGURE 7.5
(a) Pulse position peak detecting multiplier type I. (b) Pulse position peak detecting multiplier type II. (c) Pulse position peak detecting multiplier type II.

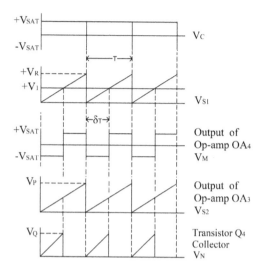

FIGURE 7.6
Associated waveforms of Figure 7.5.

A positive going ramp V_{S1} is generated at the output of op-amp OA_1. When the output of OA_1 reaches the voltage level of $+V_R$, the comparator OA_2 output becomes LOW. The transistor Q_1 is ON and hence the capacitor C_1 is shorted so that op-amp OA_1 output becomes ZERO. Then op-amp OA_2 output goes to HIGH, transistor Q_1 is OFF and the integrator composed by R_1, C_1 and op-amp OA_1 integrates the reference voltage $-V_R$ and the cycle therefore repeats to provide (1) a saw tooth wave of peak value V_R at the output of op-amp OA_1 and (2) a short pulse waveform V_C at the output of comparator OA_2. The short pulse V_C also controls transistor Q_2. During the short LOW time of V_C, transistor Q_2 is ON, the capacitor C_2 is short-circuited so that op-amp OA_3 output is zero volts. During HIGH time of V_C, transistor Q_2 is OFF, the integrator formed by resistor R_2, capacitor C_2 and op-amp OA_3 integrates its input voltage $-V_2$, and its output is given as

$$V_{S2} = -\frac{1}{R_2 C_2}\int -V_2\, dt = \frac{V_2}{R_2 C_2}t \qquad (7.16)$$

Another saw tooth waveform V_{S2} with peak value V_P is generated at the output of op-amp OA_3. From the waveforms shown in Figure 7.6, Equations 7.15 and 7.16, and the fact that at $t = T$, $V_{S1} = V_R$, $V_{S2} = V_P$:

$$V_R = \frac{V_R}{R_1 C_1}T$$

$$T = R_1 C_1 \qquad (7.17)$$

$$V_P = \frac{V_2}{R_2 C_2} T \tag{7.18}$$

In Figure 7.5b, as discussed in Chapter 3 (Section 3.1, Figure 3.1d), a saw tooth wave V_{S1} of peak value V_R and time period T is generated by op-amp OA_1, resistor R_1, capacitor C_1, and transistors Q_1 and Q_2.

$$V_R = 2V_{BE} \tag{7.19}$$

$$T = 2R_1 C_1 \tag{7.20}$$

The short pulse V_C is also given to transistor Q_3, which constitutes a controlled integrator along with op-amp OA_3, resistor R_2 and capacitor C_2. During HIGH value of V_C, the transistor Q_3 is OFF, and another integrator is formed by op-amp OA_3, resistor R_2 and capacitor C_2. The integrator output is given as

$$V_{S2} = -\frac{1}{R_2 C_2} \int -V_2 \, dt = \frac{V_2}{R_2 C_2} t \tag{7.21}$$

During LOW value of V_C, the transistor Q_3 is ON and hence the capacitor C_2 is short-circuited so that the integrator output becomes zero. Another saw tooth wave V_{S2} with peak value of V_P is generated at the output of integrator OA_3. From the waveforms shown in Figure 7.6, Equation 7.21, and the fact that at $t = T$, $V_{S2} = V_P$:

$$V_P = \frac{V_2}{R_2 C_2} T \tag{7.22}$$

In Figure 7.5c, the op-amp OA_1 along with transistor Q_1; resistors R_1, R_6, R_3, R_4 and R_5; and capacitor C_1 constitute a saw tooth wave generator. A saw tooth wave V_{S1} with peak value of V_R is generated by this saw tooth wave generator. The time period T of this saw tooth wave V_{S1} is given as

$$T = 2R_5 C_1 \ln\left(1 + 2\frac{R_1}{R_6}\right)$$

$$V_R = \beta(V_{CC}) + \frac{\beta(V_{CC})}{1.5}$$

where β is given as $\beta = \dfrac{R_1}{R_1 + R_6}$.

The short pulse waveform V_C from the saw tooth wave generator controls transistor Q_2. During HIGH time of V_C, transistor Q_2 is OFF, and an integrator is formed by op-amp OA_3, resistor R_2 and capacitor C_2. The integrator output is given as

$$V_{S2} = -\frac{1}{R_2 C_2} \int -V_2 \, dt = \frac{V_2}{R_2 C_2} t \qquad (7.23)$$

During the short LOW time of V_C, the transistor Q_2 is ON and the capacitor C_2 is short-circuited so that the integrator output becomes zero volts. Another saw tooth wave V_{S2} with peak value of V_P is generated at the output of op-amp OA_3.

From Equation 7.23, waveforms in Figure 7.6 and the fact that at $t = T$, $V_{S2} = V_P$:

$$V_P = \frac{V_2}{R_2 C_2} T \qquad (7.24)$$

Two saw tooth waveforms V_{S1} and V_{S2} of the same time period T are generated in all circuits of Figure 7.5.

In all Figure 7.5, the saw tooth wave V_{S1} is compared with the first input voltage V_1 by the comparator OA_4, and a rectangular wave V_M is generated at the output of comparator OA_4. The OFF time of this rectangular waveform V_M is given as

$$\delta_T = \frac{V_1}{V_R} T \qquad (7.25)$$

The rectangular pulse V_M controls transistor Q_4. During LOW time of V_M, the transistor Q_4 is OFF and the saw tooth wave V_{S2} exists at the collector of transistor Q_4. During HIGH time of V_M, the transistor Q_4 is ON and zero volts exist on the collector of transistor Q_4. A semi saw tooth wave V_N with peak value V_Q is generated at the collector of transistor Q_4. The peak value V_Q is given as

$$V_Q = \frac{V_P}{T} \delta_T \qquad (7.26)$$

The peak detector realized by op-amp OA_5, diode D_1 and capacitor C_3 gives this peak value V_Q at its output. Hence

$$V_O = V_Q$$

Equations 7.24 and 7.25 in Equation 7.26 gives

$$V_O = \frac{V_1 V_2}{V_R} \frac{T}{R_2 C_2} \qquad (7.27)$$

Let $T = R_2 C_2$:

$$V_O = \frac{V_1 V_2}{V_R} \qquad (7.28)$$

Design Exercise

Replace the transistors switches shown in Figure 7.5 with the FET switches of Figure 1.14 and the MOSFET switches of Figure 1.15 (see Chapter 1). In each, (i) draw circuit diagrams, (ii) explain their working operation, (iii) draw waveforms at appropriate places and (iv) deduce expressions for their output voltages.

7.4 Pulse Position Peak Sampling Multipliers with Transistors

The circuit diagrams of pulse position peak sampling multipliers with transistors are shown in Figure 7.7 and their associated waveforms are shown in

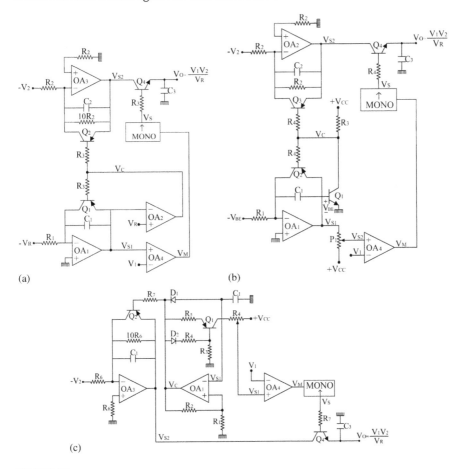

FIGURE 7.7
(a) Pulse position peak detecting multiplier type I. (b) Pulse position peak detecting multiplier type II. (c) Pulse position peak detecting multiplier type II.

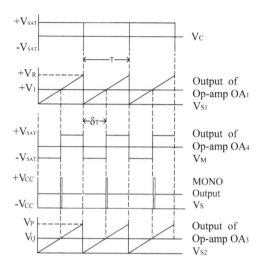

FIGURE 7.8
Associated waveforms of Figure 7.7.

Figure 7.8. The circuits in Figure 7.7 are similar to circuits shown in Figure 7.5 except at the output stage.

As discussed in Section 7.3, two saw tooth waves V_{S1} and V_{S2} of the same time period T are generated at the outputs of OA_1 and OA_3 with peak voltages of V_R and V_P, respectively.

From Equations 7.18, 7.22 and 7.24, the peak value of the second saw tooth wave V_{S2} is given as

$$V_P = \frac{V_2}{R_2C_2}T \qquad (7.29)$$

A rectangular wave V_M is generated at the output of comparator OA_4. From Equation 7.25, the OFF time of this rectangular wave V_M is given as

$$\delta_T = \frac{V_1}{V_R}T \qquad (7.30)$$

The rectangular pulse V_M is given to the monostable multivibrator, which gives a narrow spike V_S during every rising edge of V_M. This spike V_S acts as a sampling pulse to the sample and hold circuit. The sample and hold circuit realized by transistor Q_4 and capacitor C_3 gives the peak value of V_{S2} at the time of δ_T. From Figure 7.8, the sampled output is V_Q.
    The sampled value V_Q is given as

$$V_Q = \frac{V_P}{T}\delta_T \qquad (7.31)$$

$$V_O = V_Q$$

Equations 7.29 and 7.30 in Equation 7.31 give

$$V_O = \frac{V_1 V_2}{V_R} \frac{T}{R_2 C_2}$$

Let $T = R_2 C_2$, then

$$V_O = \frac{V_1 V_2}{V_R} \tag{7.32}$$

Design Exercise

Replace the transistors switches shown in Figure 7.7 with the FET switches of Figure 1.14 and the MOSFET switches of Figure 1.15 (see Chapter 1). In each, (i) draw circuit diagrams, (ii) explain their working operation, (iii) draw waveforms at appropriate places and (iv) deduce expressions for their output voltages.

Tutorial Exercises

7.1 In the multiplier circuits shown in Figure 7.5, if the polarity of input voltage V_2 and direction of diode D_1 are reversed, (i) draw waveforms at appropriate places and (ii) deduce expressions for their output voltages.

7.2 In the multiplier circuit shown in Figure 7.5a, if $-V_{CC}$ is changed to $+V_{CC}$ and V_R is changed to $-V_R$, (i) draw waveforms at appropriate places and (ii) deduce the expression for its output.

7.3 In the multiplier circuits shown in Figure 7.7, if the polarity of input voltage V_2 is reversed, (i) draw waveforms at appropriate places and (ii) deduce expressions for their output voltages.

7.4 In the multiplier circuits shown in Figure 7.3, if the polarity of input voltage V_2 and direction of diode D_1 are reversed, (i) draw waveforms at appropriate places and (ii) deduce expressions for their output voltages.

7.5 A saw tooth waveform is generated along with a short pulse of short duration from $-V_{CC}$ to $+V_{CC}$. Design a (i) pulse position detecting multiplier with multiplexer, (ii) pulse position detecting multiplier with switches, (iii) pulse position sampling multiplier with multiplexers, (iv) pulse position sampling multiplier with switches, (v) pulse position detecting multiplier with transistors and (vi) pulse position sampling multiplier with transistors.

7.6 A short pulse from $+V_{CC}$ to $-V_{CC}$ during the short time is given to you as a block. Design a (i) pulse position detecting multiplier with multiplexer, (ii) pulse position detecting multiplier with switches, (iii) pulse position sampling multiplier with multiplexers, (iv) pulse position sampling multiplier with switches, (v) pulse position detecting multiplier with transistors and (vi) pulse position sampling multiplier with transistors.

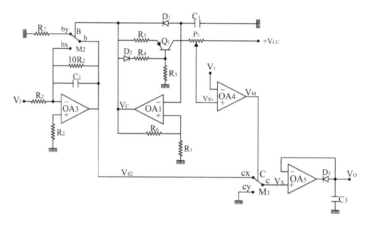

FIGURE 7.9
Circuit for Tutorial Exercise 7.7.

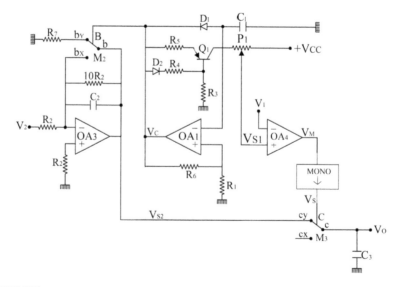

FIGURE 7.10
Circuit for Tutorial Exercise 7.8.

7.7 For the circuit shown in Figure 7.9, (i) explain the working operation, (ii) draw waveforms at appropriate places and (iii) deduce expressions for the output voltages.

7.8 For the circuit shown in Figure 7.10, (i) explain the working operation, (ii) draw waveforms at appropriate places and (iii) deduce expressions for the output voltages.

8

Peak Responding Multipliers with V/T and V/f Converters

As discussed in Chapter 1, Section 1.10, an op-amp–based astable multivibrator can be extended to perform as a voltage-to-period (V/T) converter or a voltage-to-frequency (V/f) converter. Peak detecting multipliers and peak sampling multipliers can be developed using these V/T and V/f converters and are described in this chapter.

8.1 Peak Responding Multipliers with V/T Converter: Multiplexing Type

The multiplexing type peak responding multipliers using a voltage-to-period converter are shown in Figure 8.1 and their associated waveforms are shown in Figure 8.2. Figure 8.1a shows a peak detecting multiplier and Figure 8.1b shows a peak sampling multiplier. A square wave V_C is generated by op-amp OA_1; resistors R_1, R_2, R_3; capacitor C_1; and multiplexer M_1. The square wave V_C controls multiplexer M_2. The multiplexer M_2 connects $-V_2$ during HIGH time of V_C and $+V_2$ during LOW time of V_C. Another square wave V_N with $\pm V_2$ peak-to-peak values is generated at the multiplexer M_2 output. The integrator formed by op-amp OA_2, resistors R_4 and capacitor C_2 converts the square wave V_N into a triangular wave V_{T2} of $\pm V_P$ peak-to-peak values. V_P is proportional to V_2:

$$V_P = K_1 V_2 \tag{8.1}$$

The time period T of square wave V_C is proportional to V_1:

$$T = K_2 V_1 \tag{8.2}$$

The integrator output is given as

$$V_{T2} = \frac{1}{R_4 C_2} \int V_2 \, dt = \frac{V_2}{R_4 C_2} t \tag{8.3}$$

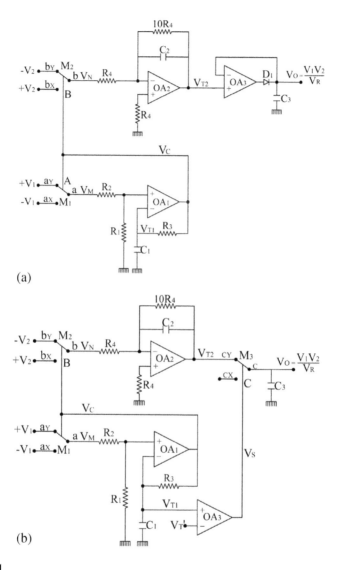

FIGURE 8.1
(a) Peak detecting multiplier using a voltage-to-period converter. (b) Peak sampling multiplier using a voltage-to-period converter.

The half time period is $t = T/2$. From the waveforms shown in Figure 8.2 and from Equation 8.3 with $t = T/2$, $V_{T2} = 2\,V_P$:

$$2V_P = \frac{V_2}{R_4C_2}\frac{T}{2}; \quad V_P = \frac{V_2}{4R_4C_2}V_1K_2 \tag{8.4}$$

K_1 and K_2 are constant values.

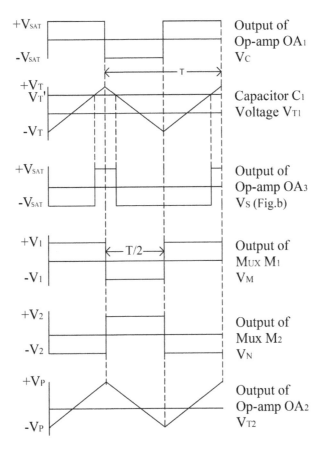

FIGURE 8.2
Associated waveforms of Figure 8.1.

Let

$$V_R = \frac{4R_4C_2}{K_2}$$

$$V_P = \frac{V_1V_2}{V_R} \tag{8.5}$$

1. In Figure 8.1a, the peak detector realized by op-amp OA_3, diode D_1 and capacitor C_3 gives the peak value V_P at the output: $V_0 = V_P$.

2. In Figure 8.1b, the peak value V_P is sampled by the sample and hold circuit realized by multiplexer M_3 and capacitor C_3 with a sampling pulse V_S. The sampling pulse V_S is generated by comparing capacitor C_1 voltage V_{T1} with slightly less than its peak value V_T, i.e., V_T'. The sampled output is $V_0 = V_P$.

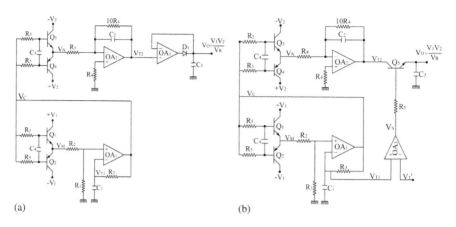

(a)   (b)

FIGURE 8.3
(a) Peak detecting multiplier using a voltage-to-period converter. (b) Peak sampling multiplier using a voltage-to-period converter.

From Equation 8.5, the output voltage is given as $V_O = V_P$:

$$V_O = \frac{V_1 V_2}{V_R} \qquad (8.6)$$

Design Exercises

1. The multiplexers M_1 and M_2 in Figure 8.1 can be replaced with the transistorized multiplexers of Figure 1.17 (see Chapter 1) and shown in Figure 8.3. (i) Explain the working operation of the multipliers shown in Figure 8.3, (ii) draw waveforms at appropriate places and (iii) deduce expressions for their outputs.

2. The multiplexers M_1 and M_2 in Figure 8.1 are to be replaced with the FET multiplexers of Figure 1.18 and the MOSFET multiplexers Figure 1.19 (see Chapter 1). In each, (i) draw the circuit diagrams, (ii) explain the working operation, (iii) draw waveforms at appropriate places and (iv) deduce expressions for the output voltage.

8.2 Peak Responding Multipliers with V/T Converter: Switching Type

The switching-type peak responding multipliers using a voltage-to-period converter are shown in Figure 8.4 and their associated waveforms are shown in Figure 8.5. Figure 8.4a shows a series switching peak detecting

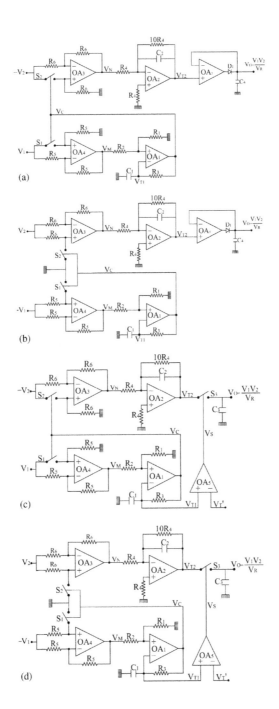

FIGURE 8.4
(a) Series switching peak detecting multiplier using a voltage-to-period converter. (b) Parallel switching peak detecting multiplier using a voltage-to-period converter. (c) Series switching sampling multiplier using a voltage-to-period converter. (d) Parallel switching sampling multiplier using a voltage-to-period converter.

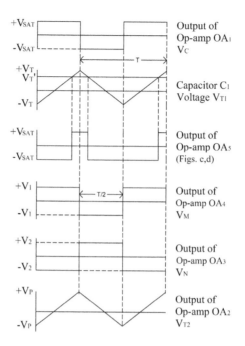

FIGURE 8.5
Associated waveforms of Figure 8.4.

multiplier, Figure 8.4b shows a shunt or parallel switching peak detecting multiplier, Figure 8.4c shows a series switching peak sampling multiplier, and Figure 8.4d shows a shunt or parallel switching peak sampling multiplier. A square wave V_C is generated by op-amp OA_1; resistors R_1, R_2, R_3; and capacitor C_1.

1. First, let us consider the circuits shown in Figure 8.4a and c. During HIGH time of the square waveform V_C, (1) the switch S_1 is closed, the op-amp OA_4 along with resistors R_5 will work as a non-inverting amplifier and $+V_1$ will appear at its output ($V_M = +V_1$); and (2) the switch S_2 is closed, the op-amp OA_3 along with resistors R_6 will work as a non-inverting amplifier and $-V_2$ will appear at its output ($V_N = -V_2$). During LOW time of the square waveform V_C, (1) the switch S_1 is opened, the op-amp OA_4 along with resistors R_5 will work as an inverting amplifier and $-V_1$ will appear at its output ($V_M = -V_1$); and (2) the switch S_2 is opened, the op-amp OA_3 along with resistors R_6 will work as an inverting amplifier and $+V_2$ will appear at its output ($V_N = +V_2$).

2. Next, let us consider the circuits shown in Figure 8.4b and d. During HIGH time of the square waveform V_C, (1) the switch S_1 is closed, the op-amp OA_4 along with resistors R_5 will work as an inverting

amplifier and $+V_1$ will appear at its output ($V_M = +V_1$); and (2) the switch S_2 is closed, the op-amp OA_3 along with resistors R_6 will work as an inverting amplifier and $-V_2$ will appear at its output ($V_N = -V_2$). During LOW time of the square waveform V_C, (1) the switch S_1 is opened, the op-amp OA_4 along with resistors R_5 will work as a non-inverting amplifier and $-V_1$ will appear at its output ($V_M = -V_1$); and (2) the switch S_2 is opened, the op-amp OA_3 along with resistors R_6 will work as a non-inverting amplifier and $+V_2$ will appear at its output ($V_N = +V_2$).

In all the circuits of Figure 8.4, two square waves—V_M with $\pm V_1$ peak-to-peak values at the output of op-amp OA_4 and V_N with $\pm V_2$ peak-to-peak at the output of op-amp OA_3—are generated from the astable clock V_C.

The integrator formed by op-amp OA_2, resistors R_4 and capacitor C_2 converts the square wave V_N into a triangular wave V_{T2} of $\pm V_P$ peak-to-peak values. V_P is proportional to V_2:

$$V_P = K_1 V_2 \tag{8.7}$$

The time period T of square wave V_S is proportional to V_1:

$$T = K_2 V_1 \tag{8.8}$$

The integrator output is given as

$$V_{T2} = \frac{1}{R_4 C_2} \int V_2 dt = \frac{V_2}{R_4 C_2} t \tag{8.9}$$

For the half time period, from the waveforms shown in Figure 8.5 and from Equation 8.9 with $t = T/2$, $V_{T2} = 2 V_P$:

$$2V_P = \frac{V_2}{R_4 C_2} \frac{T}{2}; \quad V_P = \frac{V_2}{4R_4 C_2} V_1 K_2 \tag{8.10}$$

K_1 and K_2 are constant values.
    Let

$$V_R = \frac{4R_4 C_2}{K_2}$$

$$V_P = \frac{V_1 V_2}{V_R} \tag{8.11}$$

1. In Figure 8.4a and b, the peak detector realized by op-amp OA_5, diode D_1 and capacitor C_4 gives the peak value V_P at the output: $V_0 = V_P$.

2. In Figure 8.4c and d, the peak value V_P is sampled by the sample and hold circuit realized by switch S_3 and capacitor C_3 with a sampling pulse V_S. The sampling pulse V_S is generated by comparing capacitor C_1 voltage V_{T1} with slightly less than its peak value V_T, i.e., V_T', by the comparator OA_5. The sampled output V_O is the peak value V_P.

From Equation 8.11, the output voltage will be $V_O = V_P$:

$$V_O = \frac{V_1 V_2}{V_R} \tag{8.12}$$

Design Exercises

1. The switches S_1 and S_2 in Figure 8.4a and b can be replaced with the transistorized switches of Figure 1.13 (see Chapter 1) and shown in Figure 8.6. (i) Explain the working operation of the multipliers shown in Figure 8.6, (ii) draw waveforms at appropriate places and (iii) deduce expressions for their output voltages.

2. The switches S_1 and S_2 in Figure 8.4a and b are to be replaced with the FET switches of Figure 1.14 and the MOSFET switches of Figure 1.15 (see Chapter 1). In each, (i) draw the circuit diagrams, (ii) explain their working operations, (iii) draw waveforms at appropriate places and (iv) deduce expressions for their output voltages.

8.3 Peak Responding Multipliers Using V/f Converter: Multiplexing Type

The peak responding multipliers using a voltage-to-frequency converter are shown in Figure 8.7 and their associated waveforms are shown in Figure 8.8. Figure 8.7a shows a peak detecting multiplier and Figure 8.7b shows a peak sampling multiplier. A square wave V_C is generated by op-amp OA_1; resistors R_1, R_2, R_3; capacitor C_1; and multiplexer M_1. The square wave V_C controls multiplexer M_2. The multiplexer M_2 connects $-V_O$ during HIGH time of V_C and $+V_O$ during LOW time of V_C. Another square wave V_N with $\pm V_O$ peak-to-peak values is generated at the multiplexer M_2 output. The integrator formed by op-amp OA_2, resistors R_4 and capacitor C_2 converts the square wave V_N into a triangular wave V_{T2} of $\pm V_P$ peak-to-peak values. V_P is proportional to V_O:

$$V_P = K_1 V_O \tag{8.13}$$

The time period T of square wave V_C is inversely proportional to V_1:

$$T = \frac{K_2}{V_1} \tag{8.14}$$

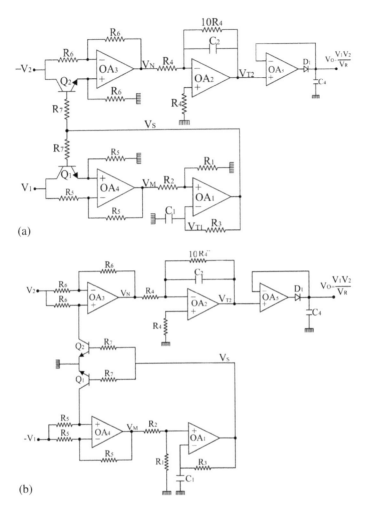

FIGURE 8.6
(a) Peak detecting multiplier using a voltage-to-period converter. (b) Shunt switching peak detecting multiplier using a voltage-to-period converter.

The integrator output is given as

$$V_{T2} = \frac{1}{R_4 C_2} \int V_O \, dt = \frac{V_O}{R_4 C_2} t \qquad (8.15)$$

For the half time period, from the waveforms shown in Figure 8.8 and from Equation 8.15 with $t = T/2$, $V_{T2} = 2V_P$:

$$2V_P = \frac{V_O}{R_4 C_2} \frac{T}{2}; \quad V_P = \frac{V_O}{4R_4 C_2} \frac{K_2}{V_1} \qquad (8.16)$$

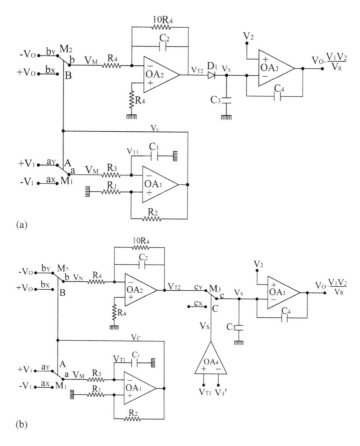

FIGURE 8.7
(a) Peak detecting multiplier using a voltage-to-frequency converter. (b) Peak sampling multiplier using a voltage-to-frequency converter.

Let

$$V_R = \frac{K_2}{4R_4C_2}$$

$$V_P = \frac{V_O V_R}{V_1} \tag{8.17}$$

1. In the circuit shown in Figure 8.7a, the peak detector realized by diode D_1 and capacitor C_3 gives the peak value V_P at the output: $V_X = V_P$.

2. In the circuit shown in Figure 8.7b, the sample and hold circuit composed by multiplexer M_3 with capacitor C_3 gives the peak value V_P by the sampling pulse V_S generated by op-amp OA_4. The sampling

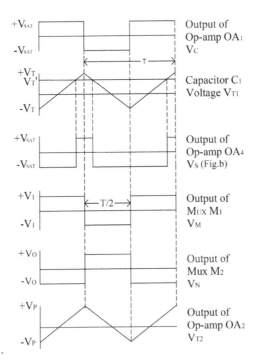

FIGURE 8.8
Associated waveforms of Figure 8.7.

pulse V_S is generated by comparing the capacitor C_1 voltage V_{T1} of $\pm V_T$ peak-to-peak value with a voltage of V_T', which is slightly less than the value of V_T by comparator OA_4. The sampled output is V_X, which is the peak value V_P: $V_X = V_P$.

From Equation 8.17

$$V_X = \frac{V_O V_R}{V_1} \tag{8.18}$$

The op-amp OA_3 is kept in a negative closed loop configuration and a positive direct current (dc) voltage is ensured in the feedback. Hence its inverting terminal voltage will equal its non-inverting terminal voltage, i.e.,

$$V_X = V_2 \tag{8.19}$$

From Equations 8.18 and 8.19:

$$V_O = \frac{V_1 V_2}{V_R} \tag{8.20}$$

Design Exercises

1. The multiplexers M_1 and M_2 in Figure 8.7 can be replaced with the transistorized multiplexers of Figure 1.7 (see Chapter 1) and shown in Figure 8.9. (i) Explain the working operation of the multipliers shown in Figure 8.9, (ii) draw waveforms at appropriate places and (iii) deduce expressions for their output voltages.

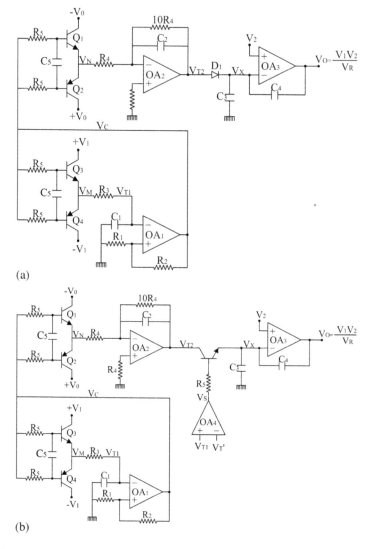

FIGURE 8.9

(a) Peak detecting multiplier using a voltage-to-frequency converter. (b) Peak sampling multiplier using a voltage-to-frequency converter.

2. The multiplexers M_1 and M_2 in Figure 8.7 are to be replaced with the FET multiplexers of Figure 1.18 and the MOSFET multiplexers of Figure 1.19 (see Chapter 1). In each, (i) draw the circuit diagram, (ii) explain the working operation, (iii) draw waveforms at appropriate places and (iv) deduce the expression for the output voltage.

8.4 Peak Responding Multipliers Using V/f Converter: Switching Type

The peak responding multipliers using a voltage-to-frequency converter are shown in Figure 8.10 and their associated waveforms are shown in Figure 8.11. Figure 8.10a shows a series switching peak detecting multiplier, Figure 8.10b shows a shunt switching peak detecting multiplier, Figure 8.10c shows a series switching peak sampling multiplier and Figure 8.10d shows a shunt switching peak sampling multiplier. A square wave V_C is generated by op-amp OA_1; resistors R_1, R_2, R_3; capacitor C_1; and a control amplifier realized by op-amp OA_3 and resistors R_6. The square wave V_C control switches S_1 and S_2.

1. First, let us first consider the circuits shown in Figure 8.10a and c. During HIGH time of the square waveform V_C, (1) the switch S_1 is closed, the op-amp OA_3 along with resistors R_6 will work as a non-inverting amplifier and $+V_1$ will appear at its output ($V_M = +V_1$); and (2) the switch S_2 is closed, the op-amp OA_4 along with resistors R_5 will work as a non-inverting amplifier and $-V_O$ will appear at its output ($V_N = -V_O$). During LOW time of the square waveform V_C, (1) the switch S_1 is opened, the op-amp OA_3 along with resistors R_6 will work as a inverting amplifier and $-V_1$ will appear at its output ($V_M = -V_1$); and (2) the switch S_2 is opened, the op-amp OA_4 along with resistors R_5 will work as an inverting amplifier and $+V_O$ will appear at its output ($V_N = +V_O$).

2. Next, let us consider the second circuit shown in Figure 8.10b and d. During HIGH time of the square waveform V_C, (1) the switch S_1 is closed, the op-amp OA_3 along with resistors R_6 will work as an inverting amplifier and $+V_1$ will appear at its output ($V_M = +V_1$); and (2) the switch S_2 is closed, the op-amp OA_4 along with resistors R_5 will work as inverting amplifier and $-V_O$ will appear at its output ($V_N = -V_O$). During LOW time of the square waveform V_C, (1) the switch S_1 is opened, the op-amp OA_3 along with resistors R_6 will work as a non-inverting amplifier and $-V_1$ will appear at its output ($V_M = -V_1$); and (2) the switch S_2 is opened, the op-amp OA_4 along with resistors R_5 will work as a non-inverting amplifier and $+V_O$ will appear at its output ($V_N = +V_O$).

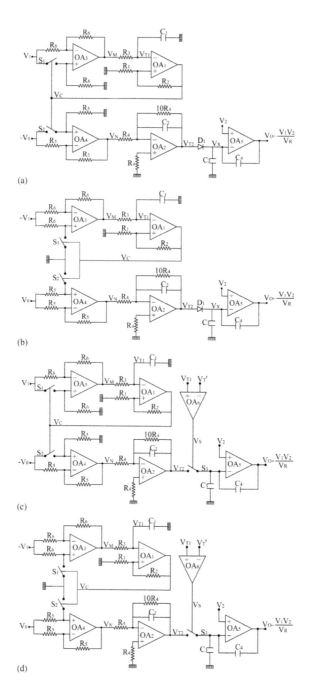

FIGURE 8.10
(a) Series switching peak detecting multiplier using a V/f converter. (b) Parallel switching peak detecting multiplier using a V/f converter. (c) Series switching sampling multiplier using a V/f converter. (d) Shunt switching sampling multiplier using a V/f converter.

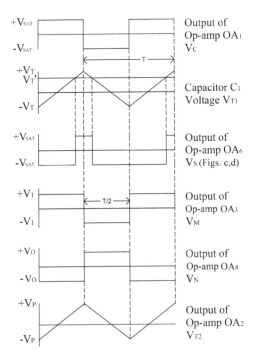

FIGURE 8.11
Associated waveforms of Figure 8.10.

In all the circuits of Figure 8.10, two square waves—V_M with $\pm V_1$ peak-to-peak values at the output of op-amp OA_3 and V_N with $\pm V_O$ peak-to-peak values at the output of op-amp OA_4—are generated from astable clock V_C.

The integrator formed by op-amp OA_2, resistors R_4 and capacitor C_2 converts the square wave V_N into a triangular wave V_{T2} of $\pm V_P$ peak-to-peak values. V_P is proportional to V_O:

$$V_P = K_1 V_O \tag{8.21}$$

The time period T of square wave V_C is inversely proportional to V_1:

$$T = \frac{K_2}{V_1} \tag{8.22}$$

The integrator output is given as

$$V_{T2} = \frac{1}{R_4 C_2} \int V_O \, dt = \frac{V_O}{R_4 C_2} t \tag{8.23}$$

For the half time period, from the waveforms shown in Figure 8.11 and from Equation 8.23 with $t = T/2$, $V_{T2} = 2 V_P$:

$$2V_P = \frac{V_O}{R_4C_2}\frac{T}{2}; \quad V_P = \frac{V_O}{4R_4C_2}\frac{K_2}{V_1}$$

K_1 and K_2 are constant values.
Let

$$V_R = \frac{K_2}{4R_4C_2}$$

$$V_P = \frac{V_O}{V_1}V_R \tag{8.24}$$

1. In Figure 8.10a and b the peak detector realized by diode D_1 and capacitor C_4 gives the peak value V_P at the output: $V_X = V_P$.
2. In Figure 8.10c and d, the peak value V_P is sampled by the sample and hold circuit realized by switch S_3 and capacitor C_3 with a sampling pulse V_S. The sampling pulse V_S is generated by comparing capacitor C_1 voltage V_{T1} with slightly less than its peak value V_T, i.e., V_T', by the comparator OA_6. The sampled output V_X is the peak value V_P.

The op-amp OA_5 is kept in a negative closed loop feedback configuration and a positive dc voltage is ensured in the feedback loop. Hence

$$V_X = V_2 \tag{8.25}$$

From Equations 8.24 and 8.25

$$V_O = \frac{V_1 V_2}{V_R} \tag{8.26}$$

Design Exercises

1. The switches in Figure 8.10 can be replaced with the transistorized switches of Figure 1.13 (see Chapter 1) and shown in Figure 8.12. (i) Explain the working operation of the multipliers shown in Figure 8.12, (ii) draw waveforms at appropriate places and (iii) deduce expressions for their output voltages.
2. The switches in Figure 8.10 are to be replaced with the FET switches of Figure 1.14 and the MOSFET switches of Figure 1.15 (see Chapter 1). In each, (i) draw the circuit diagrams, (ii) explain their working operations, (iii) draw waveforms at appropriate places and (iv) deduce expressions for their output voltages.

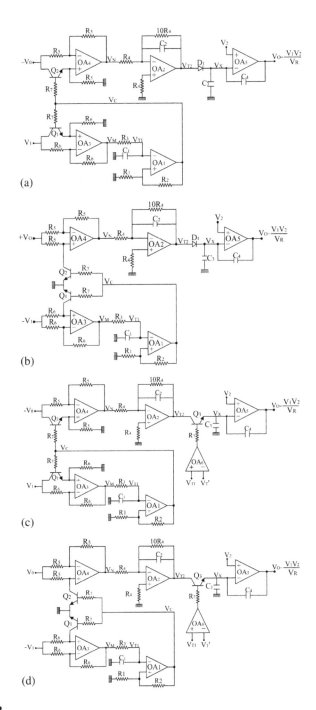

FIGURE 8.12
(a) Series switching peak detecting multiplier using a V/f converter. (b) Shunt switching peak detecting multiplier using a V/f converter. (c) Series switching peak sampling multiplier using a V/f converter. (d) Shunt switching peak sampling multiplier using a V/f converter.

Tutorial Exercises

8.1 In Figure 8.1a, if input terminals of the multiplexer are interchanged, what will be the output voltage?

8.2 In Figure 8.3a, if the diode D_1 is connected in the reverse direction, what will be the output voltage?

8.3 A multiplier circuit using a voltage-to-period converter is constructed. If the voltage-to-period converter is replaced with a voltage-to-frequency converter, what will be the output voltage?

8.4 In a shunt switching sampling multiplier using a voltage-to-frequency converter as shown in Figure 8.10d, if the voltage-to-frequency converter designed around op-amp OA_1 is replaced with a voltage-to-period converter designed around op-amp OA_1 in Figure 8.7b, what will be output voltage?

8.5 In the multiplier circuits shown in Figure 8.1, if the inputs of multiplexer M_2 are interchanged and the direction of diode D_1 is reversed, (i) draw waveforms at appropriate places and (ii) deduce expressions for their output voltages.

8.6 In the multiplier circuits shown in Figure 8.4, if the polarity of input voltage V_2 and direction of diode D_1 are reversed, (i) draw waveforms at appropriate places and (ii) deduce expressions for their output voltages.

8.7 In the multiplier circuits shown in Figure 8.7, if the inputs of multiplexer M_2 are interchanged and the direction of diode D_1 is reversed, (i) draw waveforms at appropriate places and (ii) deduce expressions for their output voltages.

8.8 In the multiplier circuits shown in Figure 8.10, if the polarity of input voltage V_2, direction of diode D_1 and polarity of V_O given to control amplifier OA_4 are reversed, (i) draw waveforms at appropriate places and (ii) deduce expressions for their output voltages.

Conclusion

The accuracy of the multiplier circuits shown in this book depend on the sharpness and linearity of the saw tooth waveform/triangular waveform. The offset voltage of all the op-amps is to be nulled for better performance of the circuits. Even though the output voltage is proportional to the input voltages, in practice, it is also proportional to the (1) attenuation constant of the low pass filter, (2) speed recovery time of diode D_1, (3) short ON/OFF time of V_C, (4) short ON time of the sampling pulse V_S and (5) slope of falling edge of saw tooth waveform V_{S1}. The maximum value of V_R or V_T can be 0.75 V_{CC}. V_1 should have a maximum value of V_R or V_T. V_2 should have a maximum value of 0.75 V_{CC}.

In Figure 3.1a–c, the polarity of V_1 should only be positive and the polarity of V_2 can be positive or negative. Hence these multipliers are the two-quadrant type.

In Figures 3.3a–c, 3.5a–c, 3.6a and b, and 3.8a and b, the polarities of V_1 and V_2 can be positive or negative. Hence these multipliers are the four-quadrant type.

In Figures 3.9a and b and 3.11a and b, the V_1 and V_2 should only be positive. Hence these multipliers are the single-quadrant type.

In Figures 4.1a, 4.3a and b, 4.5a and b, 4.6a, and 4.9a and b, the V_1 and V_2 should be single polarity only. Hence these multipliers are the single-quadrant type.

In Figures 4.1b, 4.6b, and 4.9c and d, the polarity of V_1 should only be positive and the polarity of V_2 can be positive or negative. Hence these multipliers are the two-quadrant type.

In Figures 5.1a and b, and 5.3a and b, V_1 should have positive polarity only and V_2 may have any polarity. Hence these multipliers are the two-quadrant type.

In Figures 5.4a and b, and 5.6a and b, V_1 should have positive polarity only and V_2 may have any polarity. Hence these multipliers are the two-quadrant type.

In Figures 5.7a and b, 5.9a and b, 5.10a and b, 5.12a and b, 5.13a–d, and 5.15a–d, V_1 and V_2 may have any polarity and hence these multipliers are the four-quadrant type.

In Figures 5.16a–d and 5.18a–d, V_1 and V_2 must have single polarity only and hence these multipliers are the single-quadrant type.

In Figures 6.1a and 6.3a, V_1 must have only positive polarity and V_2 must have negative polarity only. Hence these multipliers are of single-quadrant type. In Figures 6.1b and 6.3b, V_1 must have only positive polarity and V_2 may have any polarity. Hence these multipliers are the two-quadrant type.

In Figures 6.4a–d and 6.6a and b, V_1 and V_2 must have single polarity only. Hence these multipliers are the single-quadrant type.

In Figures 6.7a and b and 6.9a and b, V_1 must have only positive polarity and V_2 must have negative polarity only. Hence these multipliers are the single-quadrant type. In Figure 6.7c and d, V_1 must have only positive polarity and V_2 may have any polarity. Hence these multipliers are the two-quadrant type.

In Figures 6.10a and 6.12a, V_1 must have only positive polarity and V_2 must have negative polarity only. Hence these multipliers are the single-quadrant type. In Figures 6.10b and 6.12b, V_1 must have only positive polarity and V_2 may have any polarity. Hence these multipliers are the two-quadrant type.

In Figure 7.1a–c, the polarity of V_1 should only be positive and the polarity of V_2 should only be negative. Hence these are the single-quadrant type.

In Figure 7.1d–f, the polarity of V_1 should only be positive and V_2 may have any polarity. Hence these multipliers are the two-quadrant type.

In Figure 7.3a, the polarity of V_1 should only be positive and the polarity of V_2 should only be negative. Hence this is a single-quadrant type. In Figure 7.3b, the polarity of V_1 should only be positive and V_2 may have any polarity. Hence this is a two-quadrant type.

In Figure 7.5a–c, the polarity of V_1 should only be positive and the polarity of V_2 should only be negative. Hence these are the single-quadrant type.

In Figure 7.7a–c, the polarity of V_1 should only be positive and V_2 may have any polarity. Hence these multipliers are the two-quadrant type.

In Figures 8.1a and 8.3a, V_1 must have only positive polarity and V_2 must have negative polarity only. Hence these multipliers are the single-quadrant type. In Figures 8.1b and 8.3b, V_1 must have only positive polarity and V_2 may have any polarity. Hence these multipliers are the two-quadrant type.

In Figures 8.4a and b and 8.6a and b, V_1 must have only positive polarity and V_2 must have negative polarity only. Hence these multipliers are the single-quadrant type. In Figure 8.4c and d, V_1 must have only positive polarity and V_2 may have any polarity. Hence this multiplier is the two-quadrant type.

In Figures 8.7a and b and 8.9a and b, V_1 and V_2 should have single polarity only. Hence these multipliers are the single-quadrant type.

In Figures 8.10a–d and 8.12a–d, V_1 must have only positive polarity and V_2 must have negative polarity only. Hence these multipliers are the single-quadrant type.

Recommended Component Values

Recommended power supply: $\pm V_{CC} = \pm 7.5$ V

All op-amps OA_1–OA_6: LF356

    Multiplexers M_1–M_3: CD4053

    Switches S_1–S_4: CD4066

    Ex-OR gate: CD4030

    NPN transistors: BC547

    PNP transistors: BC557

    MONO: CD4528

    Diodes D_1–D_2: IN4148

Figure 3.1a: $R_1 = 1$ MΩ, $R_2 = R_6 = R_3 = P_1 = 10$ K, $C_1 = 39$ pF, $C_2 = 0.1$ μF

Figure 3.1b: $R_1 = 220$ KΩ, $R_6 = R_2 = 10$ K, $C_1 = 39$ pF, $C_2 = 0.1$ μF

Figure 3.1c: $R_1 = 33$ KΩ, $R_2 = R_3 = 10$ KΩ, $R_4 = 5.6$ KΩ, $R_5 = 1$ KΩ, $R_6 = 10$ KΩ, $C_1 = C_2 = 0.1$ μF

Figure 3.1d: $R_1 = 1$ MΩ, $R_2 = 1$ KΩ, $R_3 = P_1 = 10$ KΩ, $R_6 = 10$ KΩ, $C_1 = 39$ pF, $C_2 = 100$ nF

Figure 3.3a: $R_1 = 33$ K, $R_2 = R_3 = 10$ KΩ, $R_4 = 5.6$ KΩ, $R_5 = 2.5$ KΩ, $R_6 = 100$ KΩ, $C_1 = C_2 = 0.1$ μF

Figure 3.3b: $R_1 = 1$ KΩ, $R_2 = 13$ KΩ, $R_3 = 15$ KΩ, $R_6 = 100$ KΩ, $C_1 = C_2 = 0.1$ μF

Figure 3.5a: $R_1 = 33$ K, $R_2 = R_3 = 10$ KΩ, $R_4 = 5.6$ KΩ, $R_5 = 2.5$ KΩ, $R_6 = 100$ KΩ, $R_7 = 10$ K, $C_1 = C_2 = 0.1$ μF, $C_3 = 100$ pF

Figure 3.5b: $R_1 = 1$ KΩ, $R_2 = 13$ KΩ, $R_3 = 15$ KΩ, $R_6 = 100$ KΩ, $R_7 = 10$ KΩ, $C_1 = C_2 = 0.1$ μF, $C_3 = 100$ pF

Figure 3.6a and b: $R_1 = 1$ KΩ, $R_2 = 13$ KΩ, $R_3 = 15$ KΩ, $R_4 = 100$ KΩ, $C_1 = C_2 = 0.1$ μF

Figure 3.8a and b: $R_1 = 1$ KΩ, $R_2 = 13$ KΩ, $R_3 = 15$ KΩ, $R_4 = 100$ KΩ, $R_5 = 10$ KΩ, $C_1 = C_2 = 0.1$ μF, $C_3 = 100$ pF

Figure 3.9a and b: $R_1 = 1$ KΩ, $R_2 = 13$ KΩ, $R_3 = 15$ KΩ, $R_4 = 100$ KΩ, $C_1 = C_2 = C_3 = 0.1$ μF

Figure 3.11a and b: $R_1 = 1$ KΩ, $R_2 = 13$ KΩ, $R_3 = 15$ KΩ, $R_4 = 100$ KΩ, $R_5 = 10$ KΩ, $C_1 = C_2 = C_3 = 0.1$ μF, $C_4 = 100$ pF

Figure 4.1a and b: $R_1 = R_2 = 1$ MΩ, $R_3 = 10$ KΩ, $C_1 = C_2 = 39$ pF, $C_3 = 100$ nF

Figure 4.3a and b: $R_1 = R_4 = 1$ MΩ, $R_2 = 13$ KΩ, $R_3 = 15$ KΩ, $C_1 = C_2 = 39$ pF, $C_3 = C_4 = 100$ nF

Figure 4.5a and b: $R_1 = R_4 = 1$ MΩ, $R_2 = 13$ KΩ, $R_3 = 15$ KΩ, $R_5 = 10$ KΩ, $C_1 = C_2 = 39$ pF, $C_3 = C_4 = 100$ nF, $C_5 = 100$ pF

Figure 4.6a and b: $R_1 = R_4 = 1$ MΩ, $R_2 = 13$ KΩ, $R_3 = 15$ KΩ, $C_1 = C_2 = 39$ pF, $C_3 = 100$ nF

Figure 4.8a and b: $R_1 = R_4 = 1$ MΩ, $R_2 = 13$ KΩ, $R_3 = 15$ KΩ, $R_5 = 10$ KΩ, $C_1 = C_2 = 39$ pF, $C_3 = 100$ nF, $C_4 = 100$ pF

Figure 4.9a and b: $R_1 = R_2 = 1$ MΩ, $R_3 = 10$ KΩ, $R_4 = 1$ K, $C_1 = C_2 = 39$ pF, $C_3 = 100$ nF

Figure 4.9c and d: $R_1 = R_2 = 1$ MΩ, $R_3 = 10$ KΩ, $C_1 = C_2 = 39$ pF, $C_3 = 100$ nF

Figure 5.1a and b: $R_1 = 33$ KΩ, $R_2 = R_3 = P_1 = 10$ K, $R_4 = 5.6$ KΩ, $R_6 = 1$ KΩ, $R_7 = 10$ KΩ, $C_1 = C_2 = 0.1$ μF

Figure 5.3a and b: $R_1 = 33$ KΩ, $R_2 = R_3 = P_1 = 10$ K, $R_4 = 5.6$ K, $R_6 = 1$ KΩ, $R_7 = 10$ KΩ, $R_8 = 10$ KΩ, $C_1 = C_2 = 0.1$ μF

Figure 5.4a and b: $R_1 = 220$ KΩ, $R_2 = 1$ K, $R_3 = 10$ K, $C_1 = 39$ pF, $C_2 = 100$ nF

Figure 5.6a and b: $R_1 = 220$ KΩ, $R_2 = 1$ K, $R_3 = 10$ K, $R_4 = 10$ KΩ, $C_1 = 39$ pF, $C_2 = 100$ nF

Figure 5.7a and b: $R_1 = 33$ K, $R_2 = R_3 = 10$ K, $R_4 = 5.6$ K, $R_5 = 2.5$ K, $R_6 = 10$ KΩ, $R_7 = 100$ KΩ, $C_1 = C_2 = 100$ nF

Figure 5.9a and b: $R_1 = 33$ K, $R_2 = R_3 = 10$ K, $R_4 = 5.6$ K, $R_5 = 2.5$ K, $R_6 = 10$ KΩ, $R_7 = 100$ KΩ, $R_8 = 10$ KΩ, $C_1 = C_2 = 100$ nF

Figure 5.10a and b: $R_1 = 1$ KΩ, $R_2 = 13$ K, $R_3 = 15$ KΩ, $R_4 = 10$ KΩ, $R_5 = 100$ KΩ, $C_1 = C_2 = 0.1$ μF

Figure 5.12a and b: $R_1 = 1$ KΩ, $R_2 = 13$ K, $R_3 = 15$ KΩ, $R_4 = 10$ KΩ, $R_5 = 100$ KΩ, $R_6 = 10$ KΩ, $C_1 = C_2 = 0.1$ μF

Figure 5.13a–d: $R_1 = 1$ KΩ, $R_2 = 13$ K, $R_3 = 15$ KΩ, $R_4 = 10$ KΩ, $R_5 = 100$ KΩ, $C_1 = C_2 = 0.1$ μF

Figure 5.15a–d: $R_1 = 1$ KΩ, $R_2 = 13$ K, $R_3 = 15$ KΩ, $R_4 = R_6 = 10$ KΩ, $R_5 = 100$ KΩ, $C_1 = C_2 = 0.1$ μF

Figure 5.16a–d: $R_1 = 1$ KΩ, $R_2 = 13$ KΩ, $R_3 = 15$ KΩ, $R_4 = 10$ KΩ, $R_5 = 100$ KΩ, $C_1 = C_2 = 0.1$ μF, $C_3 = 2.2$ μF

Figure 5.18a–d: $R_1 = 1$ KΩ, $R_2 = 13$ KΩ, $R_3 = 15$ KΩ, $R_5 = 100$ KΩ, $R_4 = R_6 = 10$ K, $C_1 = C_2 = 0.1$ μF, $C_3 = 2.2$ μF

Figure 6.1a and b: $R_1 = R_2 = 1$ MΩ, $C_1 = C_2 = 39$ pF, $C_3 = 100$ nF

Figure 6.3a and b: $R_1 = R_2 = 1$ MΩ, $R_3 = 10$ KΩ, $C_1 = C_2 = 39$ pF, $C_3 = 100$ nF

Figure 6.4a–d: $R_1 = R_4 = 1$ MΩ, $R_2 = 13$ K, $R_3 = 15$ KΩ, $R_5 = R_6 = 10$ KΩ, $C_1 = C_2 = 39$ pF, $C_3 = C_4 = 100$ nF

Figure 6.6a and b: $R_1 = R_4 = 1$ MΩ, $R_2 = 13$ K, $R_3 = 15$ KΩ, $R_5 = R_6 = 10$ KΩ, $C_1 = C_2 = 39$ pF, $C_3 = C_4 = 100$ nF

Figure 6.7a–d: $R_1 = R_4 = 1$ MΩ, $R_2 = 13$ K, $R_3 = 15$ KΩ, $R_5 = R_6 = 10$ KΩ, $C_1 = C_2 = 39$ pF, $C_3 = 100$ nF

Figure 6.9a and b: $R_1 = R_4 = 1$ MΩ, $R_2 = 13$ K, $R_3 = 15$ KΩ, $R_5 = R_6 = R_7 = 10$ KΩ, $C_1 = C_2 = 39$ pF, $C_3 = 100$ nF

Figure 6.10a and b: $R_1 = R_2 = 1$ MΩ, $C_1 = C_2 = 39$ pF, $C_3 = 100$ nF

Figure 6.12a and b: Rd = 10 K, Cd = 1 nF, $R_1 = R_2 = 1$ MΩ, $R_3 = 10$ KΩ, $C_1 = C_2 = 39$ pF, $C_3 = 100$ nF

Figure 7.1a and d: $R_1 = 720$ KΩ, $R_2 = 1$ MΩ, $R_3 = P_1 = 10$ KΩ, $R_4 = 1$ KΩ, $C_1 = C_2 = 39$ pF, $C_3 = 100$ nF

Figure 7.1b and e: $R_1 = R_2 = 1$ MΩ, $R_3 = 10$ KΩ, $C_1 = C_2 = 39$ pF, $C_3 = 100$ nF

Figure 7.1c and f: $R_1 = 33$ K, $R_6 = R_3 = 10$ KΩ, $R_4 = 5.6$ KΩ, $R_5 = 1$ K, $R_2 = 3$ KΩ, $C_1 = C_2 = C_3 = 100$ nF

Figure 7.3a and b: $R_1 = R_2 = 1$ MΩ, $C_1 = C_2 = 39$ pF, $C_3 = 100$ nF

Figure 7.5a: $R_1 = R_2 = 1$ MΩ, $R_3 = 1$ KΩ, $R_4 = Rd = 10$ KΩ, $C_1 = C_2 = 39$ pF, $C_3 = 100$ nF, Cd = 1 nF

Figure 7.5b: $R_1 = 500$ KΩ, $R_2 = 1$ MΩ, $R_3 = R_4 = 10$ KΩ, $R_6 = 1$ KΩ, $C_1 = C_2 = 39$ pF, $C_3 = 100$ nF

Figure 7.5c: $R_1 = 33$ K, $R_3 = R_6 = R_7 = 10$ KΩ, $R_4 = 5.6$ KΩ, $R_5 = R_8 = 1$ KΩ, $R_2 = 3$ KΩ, $C_1 = C_2 = C_3 = 100$ nF

Figure 7.7a: $R_1 = R_2 = 1$ MΩ, $R_3 = Rd = 10$ KΩ, $C_1 = C_2 = 39$ pF, Cd = 1 nF

Figure 7.7b: $R_1 = 500$ KΩ, $R_2 = 1$ MΩ, $R_3 = 1$ KΩ, $R_4 = 10$ KΩ, $C_1 = C_2 = 39$ pF

Figure 7.7c: $R_1 = 33$ KΩ, $R_2 = R_3 = 10$ KΩ, $R_4 = 5.6$ KΩ, $R_5 = 1$ KΩ, $R_6 = 3$ KΩ,

Figure 8.1a and b: $R_1 = 1$ KΩ, $R_2 = R_3 = 10$ KΩ, $R_4 = 1$ KΩ, $C_1 = C_2 = C_3 = 100$ nF

Figure 8.3a and b: $R_1 = 1$ KΩ, $R_2 = R_3 = 10$ KΩ, $R_4 = 1$ KΩ, $R_5 = 10$ K, $C_1 = C_2 = C_3 = 100$ nF, $C_4 = 100$ pF

Figure 8.4a–d: $R_1 = 1$ KΩ, $R_2 = R_3 = 10$ KΩ, $R_4 = 1$ KΩ, $R_5 = R_6 = 10$ KΩ, $C_1 = C_2 = C_3 = 100$ nF

Figure 8.6a and b: $R_1 = 1$ KΩ, $R_2 = R_3 = 10$ KΩ, $R_4 = 1$ KΩ, $R_5 = R_6 = R_7 = 10$ KΩ, $C_1 = C_2 = C_3 = 100$ nF

Figure 8.7a and b: $R_1 = 1$ KΩ, $R_2 = R_3 = 10$ KΩ, $R_4 = 1$ KΩ, $C_1 = C_2 = C_3 = C_4 = 100$ nF

Figure 8.9a and b: $R_1 = 1$ KΩ, $R_2 = R_3 = 10$ KΩ, $R_4 = 1$ KΩ, $R_5 = 10$ KΩ, $C_1 = C_2 = C_3 = C_4 = 100$ nF, $C_5 = 100$ pF

Figure 8.10a–d: $R_1 = 1$ KΩ, $R_2 = R_3 = R_5 = R_6 = 10$ KΩ, $R_4 = 1$ KΩ, $C_1 = C_2 = C_3 = 100$ nF

Figure 8.12a–d: $R_1 = 1$ KΩ, $R_2 = R_3 = 10$ KΩ, $R_4 = 1$ KΩ, $R_5 = R_6 = R_7 = 10$ KΩ, $C_1 = C_2 = C_3 = C_4 = 100$ nF

Pin Details of Integrated Circuits (ICs)

The pin details of the operational amplifier LF356 IC are shown in Figure 1, analog multiplexer CD4053 IC in Figure 2, analog switches CD4066 IC in Figure 3, transistor BC547/557 in Figure 4, CD4528 MONO shown in Figure 5, and CD 4030 Ex-OR gate is shown in Figure 6.

LF 356 IC

FIGURE 1
Pin details of LF356 IC.

CD 4053 IC

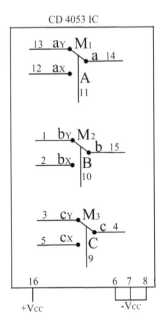

FIGURE 2
Pin details of CD4053 IC.

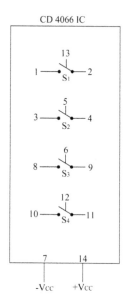

FIGURE 3
Pin details of CD4066 IC.

FIGURE 4
Pin details of BC547/BC557.

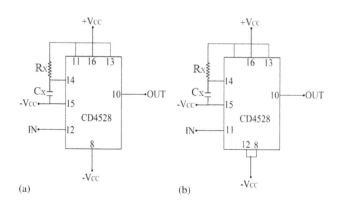

FIGURE 5
(a) Pin details of CD4528 MONO (rising edge triggering). (b) Pin details of CD4528 MONO (falling edge triggering).

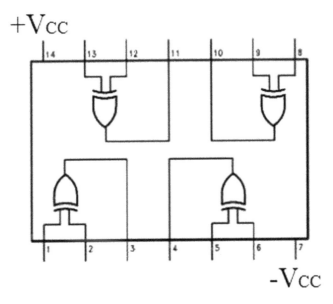

FIGURE 6
Pin details of CD4030 EX-OR Gate.

Bibliography

N. Bajaj and Jivesh Govil, "Realisation of a differential multiplier-divider based on current feedback amplifiers," *2006 IEEE International Conference on Semiconductor Electronics*, 2006, pp 708–712.

George Clayton and Steve Winder, *Operational Amplifiers*, 5th edition, Elsevier , 2003, pp 306–307.

Sergio Franco, *Design with Operational Amplifiers and Analog Integrated Circuits*, Tata McGraw-Hill Edition, 2002.

Weixin Gai, Hongyi Chen and E. Seevinck, "Quadratic-translinear CMOS multiplier-divider circuit," *Electronics Letters*, Vol. 33, No. 10, May 1997, pp 860–861.

G. Han and E. Sanchez-Sinencio, "CMOS transconductance multipliers: A tutorial," *IEEE CAS II: Analog and Digital Signal Processing*, Vol. 45, No. 12, Dec 1998, pp 1550–1563.

Greg Johnson, "Analysis of modified Tomata-Sugiyama-Yamaguchi multiplier," *IEEE Transactions on Instrumentation and Measurement*, Vol. 33, No. 1, 1984, pp 11–16.

A.J. Peyton and V. Walsh, *Analog Electronics with Op-Amps*, Cambridge University Press, 1993.

M.S. Piedade, "New analogue multiplier-divider circuit based on cyclic data convertors," *Electronics Letters*, Vol. 26, No. 1, January 1990, pp 2–4.

T.S. Rathore and B.B. Bhatacharya, "A novel type of analog multiplier," *IEEE Transactions on Industrial Electronics*, Vol. IE-31, 1984, pp 268–271.

D. Roy Choudhury and Shail B. Jain, *Linear Integrated Circuits*, 3rd edition, New Age International Publishers, 2010, pp 220–222.

C. Selvam, "A simple and low cost pulse time multiplier," *IETE Students Journal*, Vol. 34, Nos. 3 and 4, July–Dec 1993, pp 207–211.

C. Selvam and V. Jagadeesh Kumar, "A simple and inexpensive implementation of time division multiplier for two quadrant operation," *IETE Technical Review*, Vol. 12. No. 1, Jan–Feb 1995, pp 33–35.

C. Selvam and V. Jagadeesh Kumar, "A simple multiplier and squarer circuit," *IETE Students Journal*, Vol. 37. Nos. 1 and 2, Jan–Mar 1996, pp 7–10.

K.C. Selvam, "A double single slope multiplier cum divider," *IETE Journal of Education*, Vol. 41, Nos. 1 and 2, Jan–June 2000, pp 3–5.

K.C. Selvam, "A pulse position sampled multiplier," *IETE Journal of Education*, Vol. 54, No. 1, Jan–June 2013, pp 5–8.

K.C. Selvam, "Double dual slope multiplier-cum-divider," *Electronics Letters*, Vol. 49, No. 23, November 2013, pp 1435–1436.

K.C. Selvam, "Some techniques of analog multiplication using op-amp based sigma generator," *IETE Journal of Education*, Vol. 55, No. 1, Jan–June 2014, pp 33–39.

K.C. Selvam, "Four quadrant time division multiplier without using any reference clock," *IETE Journal of Education*, Vol. 58, No. 2, November 2017, pp 78–82.

M. Tomata, Y. Sugiyamma and K. Yamaguchi, "An electronic multiplier for accurate power measurement," *IEEE Transactions on Instrumentation and Measurement*, Vol. IM-17, 1968, pp 245–251.

Yu Jen Wong and Williams E. Ott, *Function Circuits: Design and Applications*, McGraw Hill Book Company, 1976.

Index